# Handbook of Microbiological Quality Control

**Pharmaceuticals and Medical Devices**

# New and Forthcoming Titles in the Pharmaceutical Sciences

# Handbook of Microbiological Quality Control

## Pharmaceuticals and Medical Devices

Edited by

**ROSAMUND M. BAIRD**

*Department of Pharmacy and Pharmacology, University of Bath, UK*

**NORMAN A. HODGES**

*School of Pharmacy and Biomolecular Sciences, University of Brighton, UK*

**STEPHEN P. DENYER**

*School of Pharmacy and Biomolecular Sciences, University of Brighton, UK*

First published 2000 by Taylor & Francis
11 New Fetter Lane, London EC4P 4EE

Simultaneously published in the USA and Canada
by Taylor & Francis Inc,
29 West 35th Street, New York, NY 10001

*Taylor & Francis is an imprint of the Taylor & Francis Group*

© Taylor & Francis and Rosamund M. Baird, Norman A. Hodges, Stephen P. Denyer

Typeset in Times by Keyword Publishing Services Ltd
Printed and bound in Great Britain by TJ International Ltd, Padstow

*British Library Cataloguing in Publication Data*
A catalogue record for this book is available from the British Library

*Library of Congress Cataloging in Publication Data*
Handbook of microbiological quality control: pharmaceuticals and medical devices/
edited by Stephen Denyer, Rosamund M. Baird, Norman A. Hodges.
    p. cm.–(Taylor & Francis series in pharmaceutical sciences)
  Includes bibliographical references and index.
    1. Pharmaceutical microbiology–Handbooks, manuals, etc. 2.
Drugs–Sterilization–Handbooks, manuals, etc. 3. Pharmaceutical industry–Quality
control–Handbooks, Manuals, etc. I. Denyer, Stephen (Stephen Paul) II, Baird, R. M.
(Rosamund M.) III. Hodges, Norman A. IV. Series.

  QR46.5. H36 2000
  615'.19–dc21                                                    00-041187

ISBN 0-748-40614-X

# Contents

**9   Endotoxin Testing                                                144**

**Alan Baines**

**10   Antimicrobial Preservative Efficacy Testing                    168**

**Norman Hodges and Geoffrey Hanlon**

# Contributors

**Alan Baines** is Managing Director of BioWhittaker UK Ltd, Wokingham, Berkshire, RG41 2PL, United Kingdom

**Rosamund M. Baird** is at the Department of Pharmacy and Pharmacology, University of Bath, Claverton Down, Bath BA2 7AY, United Kingdom

**Andrew Bill** is at the Medicines Control Agency, 2nd Floor, Prudential House, 28–40 Blossom Street, York, YO24 1GJ, United Kingdom

**Geoffrey Hanlon** is at the School of Pharmacy and Biomolecular Sciences, Brighton University, BN2 4GJ, United Kingdom

**Norman Hodges** is at the School of Pharmacy and Biomolecular Sciences, University of Brighton, BN2 4GJ, United Kingdom

**Martin Lush** is at David Begg Associates, 10 High Market Place, Kirbymoorside, Yorkshire, YO6 6AX, United Kingdom

**Ronnie Millar** is at GlaxoWellcome Research and Development, Ware, Hertfordshire, SG12 0DP, United Kingdom

**Paul Newby** is at GlaxoWellcome Product Supply, Harmire Road, Barnard Castle, Co. Durham, United Kingdom

**Anthony W. Smith** is at the Department of Pharmacy and Pharmacology, University of Bath, Claverton Down, Bath, BA2 7AY, United Knigdom

**Colin Thompson** is at GlaxoWellcome, Temple Hill, Dartford, Kent, DA1 5AH, United Kingdom

# Preface

Microbiological concerns in product manufacture, regardless of whether it involves pharmaceuticals, cosmetics, toiletries or medical devices, continue to challenge the minds of those associated with their production. Whilst much is known about how contamination arises, and control methods are well understood, there is no room for complacency; every microbiologist knows that from time to time a contamination issue inevitably will arise and must be promptly addressed. Detailed practical guidance on appropriate microbiological methods is an absolute necessity if such problems are to be resolved speedily and to the satisfaction of all involved.

This handbook attempts to fulfil this role in the laboratory, bringing together a wealth of information from diverse sources into a single volume. The reader is guided throughout the book by the advice and recommendations of expert contributors with an in-depth knowledge and personal experience of the likely practical issues to be faced by those working in the microbiology laboratory. Beginning with safety concerns and the design of laboratory facilities, guidance is provided on the choice and requirements of culture media, followed by a detailed consideration of specialized microbiological methods, including chapters on sampling, enumeration and identification of micro-organisms, involving both traditional and rapid techniques, sterility testing, environmental monitoring, preservative efficacy testing, endotoxin testing and antibiotic bioassays. Whilst established reference methods are of necessity included here, there is also an abundance of information on how these tests are interpreted in common practice. Cleaning and disinfection considerations, frequently a much neglected subject, are also addressed within this volume. Hazard analysis and audit approaches complete the routine tasks facing the company microbiologist and the reader is directed in how this might be successfully achieved. Finally, a number of worked case examples have been included to assist those involved in such every day calculations.

Although the book is intended as a complete reference source in its own right, it is in fact a companion book to the *Guide to Microbiological Control in Pharmaceuticals and Medical Devices*, shortly to be published in its second edition by Taylor and Francis. The intention here was to provide in the *Handbook* detailed practical information on methodology of the techniques advocated and identified for microbiological quality control in the *Guide*. The *Handbook* is therefore extensively cross-referenced to the *Guide*, and it is hoped that the collected wisdom of these two books provides the laboratory microbiologist with a unique, valued and much consulted reference source.

Rosamund M. Baird
Norman A. Hodges
Stephen P. Denyer

# 1

# Safe Microbiological Practices

**ANTHONY W. SMITH**

*Department of Pharmacy and Pharmacology, University of Bath, Claverton Down, Bath BA2 7AY, United Kingdom*

## 1.1 Scope and aims

This chapter sets out to offer practical advice on safe microbiological practices to all persons involved with microbiological quality assurance. It describes the current legislation relevant to handling micro-organisms and discusses the key elements of safe microbial practices, namely facility design, personnel and training, and validation and monitoring of procedures.

## 1.2 Overview of current legislation

The key feature of current legislation is risk assessment, recognizing that only after all the component elements of a process have been fully analysed, can effective and safe procedures be put in place. The three major pieces of legislation affecting handling of micro-organisms are the:

- Health and Safety at Work Act 1974
- Control of Substances Hazardous to Health (COSHH) 1994
- Genetically Modified Organisms (Contained Use) Regulations 1992 (amended in 1996)

A major change in legislation arose in COSHH 1994 which, unlike COSHH 1988, stipulated that control measures minimizing risk of illness resulting from occupational exposure should be extended to biological agents as well as chemical agents. Resulting from the Biological Agents Directive 90/679/EEC, the definition of a biological agent used in COSHH 1994 is:

'any micro-organism, cell culture or human endoparasite, including any which have been genetically modified, which may cause infection, allergy, toxicity or otherwise create a hazard to human health'

Therefore, for the first time COSHH 1994 specifically stipulated the need to carry out an assessment of risks and to select and maintain measures where exposure to any micro-organism might present a hazard to the health of any person. Clearly, all micro-organisms vary in their ability to cause infection in humans. Indeed, those micro-organisms used in pharmaceutical quality assurance procedures range from organisms not associated with any human disease to those which have significant pathogenic potential.

## 1.3   The Advisory Committee on Dangerous Pathogens (ACDP)

### *1.3.1   Background*

The ACDP was set up in 1981 and was the successor to the Dangerous Pathogens Advisory Group. Its remit is to advise the Health and Safety Commission, the Health and Safety Executive (HSE) and Health and Agriculture Ministers on all aspects of hazards and risks resulting from exposure to pathogens.

### *1.3.2   Classification*

ACDP has categorized biological agents into four Hazard Groups based on the pathogenic potential to humans, hazard to employees, probability of transmission to the community, and availability of effective prophylaxis or treatment. These Hazard Groups are classified as follows:

- Hazard Group 1:     A biological agent unlikely to cause human disease.
- Hazard Group 2:     A biological agent that can cause human disease and may be a hazard to employees; it is unlikely to spread to the community, and there is usually effective prophylaxis or effective treatment available.
- Hazard Group 3:     A biological agent that can cause severe human disease and presents a serious hazard to employees; it may present a risk of spreading to the community, but there is usually effective prophylaxis or treatment available.
- Hazard Group 4:     A biological agent that causes severe human disease and is a serious hazard to employees; it is likely to spread to the community and there is usually no effective prophylaxis or treatment available.

The Approved List includes bacteria, viruses, fungi and parasites in Groups 2, 3 and 4. However, the guidance notes emphasize that it should not be assumed that absence from the list means that the agent is a Group 1 organism. Cell lines, which may become increasingly important in pharmaceutical processes, are not included in the Approved List; however, it is expected that they will be handled in laboratories equipped to conform to the containment requirements for Hazard Group 2 organisms. None of the organisms used in common pharmacopoeial tests is in Hazard Groups 3 and 4; however, a number of species are in Hazard Group 2 (Tables 1.1–1.3). This has important legal implications since notification of use and containment conditions are mandatory.

**Table 1.1** Micro-organisms used in *European Pharmacopoeia* (1997) Test for the efficacy of antimicrobial preservation

| Micro-organism | Culture collection number | Hazard Group |
|---|---|---|
| *Pseudomonas aeruginosa* | ATCC 9027, NCIMB 8626, CIP 82-118 | 2 |
| *Staphylococcus aureus* | ATCC 6538, NCTC 10788, NCIMB 9518, CIP 4.83 | 2 |
| *Candida albicans* | ATCC 10231, NCPF 3179, IP 48.72 | 2 |
| *Aspergillus niger* | ATCC 16404, IMI 149007, IP 1431.83 | 1 |
| *Escherichia coli* | ATCC 8739, NCIMB 8545, CIP 53.126 | 1 |
| *Zygosaccharomyces rouxii* | NCYC 381, IP 2021.92 | 1 |

ATCC, American Type Culture Collection (USA)
NCIMB, National Collection of Industrial and Marine Bacteria (UK)
NCTC, National Collection of Type Cultures (UK)
CIP and IP, Collection of the Institute Pasteur (France)
NCPF, National Collection of Pathogenic Fungi (UK)
IMI, Imperial Mycological Institute (now CABI Bioscience, UK)
NCYC, National Collection of Yeast Cultures (UK).

**Table 1.2** Bacteria recommended in the *European Pharmacopoeia* (1997) as biological monitors for sterilization processes

| Sterilization process | Micro-organism and culture collection number | Hazard Group |
|---|---|---|
| Steam | *Bacillus stearothermophilus* ATCC 7953, NCTC 10007, NCIMB 8157, CIP 52.81 | 1 |
| Dry heat | *Bacillus subtilis* var. *niger* ATCC 9372, NCIMB 8058, CIP 77.18 | 1 |
| Ionizing radiation | *Bacillus pumilus* ATCC 27142, NCTC 10327, NCIMB 10692, CIP 77.28 | 1 |
| Gaseous $H_2O_2$/Peracetic acid | *Bacillus stearothermophilus* (as above) | 1 |
| Formaldehyde/Ethylene oxide | *Bacillus subtilis* var. *niger* (as above) | 1 |

ATCC, American Type Culture Collection (USA)
NCIMB, National Collection of Industrial and Marine Bacteria (UK)
NCTC, National Collection of Type Cultures (UK)
CIP and IP, Collection of the Institute Pasteur (France)
NCPF, National Collection of Pathogenic Fungi (UK)
IMI, Imperial Mycological Institute (now CABI Bioscience, UK)
NCYC, National Collection of Yeast Cultures (UK).

The COSHH regulations call for notification to HSE 30 days in advance of first use of organisms in Groups 2, 3 or 4 where it is the intention to culture, propagate and store the organisms. For a new pharmaceutical Quality Assurance (QA) laboratory the range of organisms used need not be restricted to named organisms in Hazard Group 2, rather it is sufficient to notify use of all organisms in Hazard Group 2.

**Table 1.3**  Test organisms for antibiotic assays in the *United States Pharmacopoeia* (XXIII, 1995) and *European Pharmacopoeia* (1997)

| Micro-organism | Culture collection number | Hazard Group |
|---|---|---|
| *Bacillus pumilus* | NCTC 8241, CIP 76.18 | 1 |
| *Bacillus subtilis* | ATCC 6633, NCTC 10400, CIP 52.62 | 1 |
| *Bordetella bronchispetica* | ATCC 4617, NCTC 8344, CIP 53.157 | 2 |
| *Escherichia coli* | ATCC 10536, NCIB 88795, CIP 54.127 | 1 |
| *Klebsiella pneumoniae* | ATCC 10031, NCTC 7427, CIP 53.153 | 2 |
| *Micrococcus flavus (luteus)* | ATCC 10240, NCTC 7743, CIP 53.160 | 1 |
| *Mycobacterium smegmatis* | ATCC 607 | 1 |
| *Pseudomonas aeruginosa* | ATCC 25619 | 2 |
| *Saccharomyces cerevisiae* | ATCC 9763, NCYC 87, CIP 52.62 | 1 |
| *Staphylococcus epidermidis* | ATCC 12228, NCIB 8853, CIP 68.21 | 1 |
| *Staphylococcus aureus* | ATCC 29737, CIP 53156 | 2 |
| *Streptococcus faecium* | ATCC 10541 | 2 |

ATCC, American Type Culture Collection (USA)
NCIMB, National Collection of Industrial and Marine Bacteria (UK)
NCTC, National Collection of Type Cultures (UK)
CIP and IP, Collection of the Institute Pasteur (France)
NCPF, National Collection of Pathogenic Fungi (UK)
IMI, Imperial Mycological Institute (now CABI Bioscience, UK)
NCYC, National Collection of Yeast Cultures (UK).

Tables 1.1–1.3 list the commonly used organisms in pharmaceutical QA processes, together with their ACDP classification.

### 1.3.3  *Containment Level 1*

Each of the four Hazard Groups is associated with a degree of laboratory containment relevant to the risk of occupational and community exposure from organisms within the Group. Laboratory Containment Level 1 is suitable for work with agents in Group 1 and should conform to Good Laboratory Practice. The main features of a Containment Level 1 facility are described in full since there are elements pertinent to laboratory design and to establishing Local Rules.

1   The laboratory should be easy to clean. Bench surfaces should be impervious to water and resistant to acids, alkalis, solvents and disinfectants.

2   Effective disinfectants should be available for immediate use in the event of a spillage.

3   If the laboratory is mechanically ventilated, it is preferable to maintain an inward flow of air while work is in progress by extracting room air to atmosphere.

4   All procedures should be performed so as to minimize the production of aerosols.

5   The laboratory door should be closed when work is in progress.

6   Laboratory coats or gowns should be worn in the laboratory and removed when leaving the laboratory area.

7  Personal protective equipment, including protective clothing, must be stored in a well-defined place, checked and cleaned at suitable intervals and repaired or replaced if found to be detective.

8  Personal protective equipment which may be contaminated must be removed on leaving the working area, kept apart from uncontaminated clothing and decontaminated and cleaned or, if necessary, destroyed.

9  Eating, chewing, drinking, taking medication, smoking, storing food and applying cosmetics should be forbidden.

10  Mouth pipetting should be forbidden.

11  The laboratory should contain a basin or sink that can be used for hand washing.

12  Hands should be decontaminated immediately when contamination is suspected and before leaving the laboratory.

13  Bench tops should be cleaned after use.

14  Used glassware and other materials awaiting disinfection should be stored in a safe manner. Pipettes, for example, if placed in disinfectant should be totally immersed.

15  Contaminated materials whether for recycling or disposal, should be stored and transported in robust and leak-proof containers without spillage.

16  All waste material, if not to be incinerated, should be disposed off safely by other appropriate means.

17  Accidents and incidents should be immediately reported to and recorded by the person responsible for the work or other designated person.

### 1.3.4  *Containment Level 2*

Laboratory Containment Level 2 must be used for work with agents in Hazard Group 2. Since some organisms used in pharmaceutical QA procedures are in Hazard Group 2, this Containment Level will represent the minimum standard for safe handling of micro-organisms. There are some additional points which are statutory requirements to operate a laboratory at this level, over and above those described for Laboratory Containment Level 1, and in addition some control measures change from 'should' – indicating the ACDP's strong recommendation – to 'must' – indicating a requirement defined in legislation. The additional features required for a laboratory to operate at Containment Level 2 are as follows:

1  Access to the laboratory is restricted to authorized persons.

2  There must be specified disinfection procedures.

3  If the laboratory is mechanically ventilated, it *must* be maintained at an air pressure negative to atmosphere while work is in progress.

4  Bench surfaces *must* be impervious to water, easy to clean and resistant to acids, alkalis, solvents and disinfectants.

5  There must be safe storage of biological agents.

6   Laboratory procedures that give rise to infectious aerosols must be conducted in a microbiological safety cabinet or isolator, or be otherwise suitably contained.

7   The laboratory should contain a washbasin near the exit. The taps should be of a design that can be operated without being touched by the hands.

8   Hands should be decontaminated immediately when contamination is suspected, after handling infective agents, and before leaving the laboratory.

9   An autoclave for the sterilization of waste materials should be readily accessible in the same building as the laboratory, preferably in the laboratory suite.

10  Materials for autoclaving should be transported to the autoclave in robust containers without spillage.

11  There should be a means for the safe collection, storage and disposal of contaminated waste.

## 1.4   Risk assessment and Control of Substances Hazardous to Health Regulations 1994

In essence, these regulations require that suitable and sufficient assessment of risks to health be made by employers before work with chemicals or biological agents is started. Moreover, they require that appropriate measures to prevent or control exposure are in place, and that procedures exist to assure that measures are always in operation.

### 1.4.1   Assessment of risk

Assessment of risk requires a pragmatic approach to protocol design and a thorough and detailed understanding of the process under consideration. In the context of handling micro-organisms in the laboratory, the two 'worst-case' scenarios are loss and escape of containment and infection of the laboratory worker. The excellent book by Collins (1993) reviews in detail the history of laboratory-acquired infections, and the many important lessons that have been learned and applied to safe microbial handling.

### 1.4.2   Eight steps to risk assessment

- Step 1:   Define the work activity.
- Step 2:   Identify the hazards.
- Step 3:   Identify the control measures.
- Step 4:   Evaluate the risks.
- Step 5:   Maintain the control measures.
- Step 6:   Arrive at a conclusion.
- Step 7:   Record the assessment.
- Step 8:   Review the assessment.

For a pharmaceutical QA process, Step 2 is easily satisfied by reference to the tables in this chapter and to the ACDP *Categorization of biological agents according to hazard and categories of containment* (4th edition, 1995). Steps 3 and 4 are somewhat interdependent in that effective control measures cannot be considered independently of the risks involved. Step 3 covers the issue of control safeguards such as biological cabinets, personal protective equipment and personnel instruction, training and supervision (discussed later), whereas Step 4 requires evaluation of the risks, such as transmission routes and potential for exposure, which remain despite the control measures put in place.

It is interesting to note that many surveys, reviewed by Collins (1993), indicate that an infection is more likely to arise simply as a consequence of working with a micro-organism than through an accident, although clearly the risk assessment should consider both possibilities. Assessment of risk will result in prohibiting some activities and acknowledging that in some procedures a degree of risk is unavoidable. In such circumstances, the challenge of effective risk assessment is to ensure risk minimization, for example through process and facility design and monitoring and personnel training. It should be implicit in risk assessment that the process is not a 'once-only' event and that it should be modified and up-dated as new information becomes available. Much of what we may take for granted for safe practices in the microbiology laboratory, although perhaps previously developed on an *ad hoc* basis, result from risk minimization based on an awareness of the likely routes of self-contamination and infection. Examples of development of safe practices based on knowledge of the principal routes of infection are given below:

1 Through the mouth. Therefore eating, drinking, smoking and application of cosmetics in the laboratory must be prohibited as prescribed in the conditions for operating a Containment Level 1 facility. Mouth pipetting is also a significant hazard and is prohibited. Transfer of micro-organisms to the mouth (or hair) by contaminated hands is a personnel training issue.

2 Through the skin. Laboratory personnel applying cosmetics or handling micro-organisms with cuts and abrasions are clearly risks, and again represent training issues. Accidental inoculation with a 'sharp' is always a risk and so consideration should be given to substitution wherever possible.

3 Through the eye. There is evidence that some organisms, perhaps hepatitis B, could be transmitted via the eye. While this is not directly relevant to pharmaceutical QA procedures, it should always be considered when handling any unscreened blood products, such as serum samples for drug level monitoring and pharmacokinetic analysis. Of more direct concern are organisms such as *Staphylococcus aureus* and *Pseudomonas aeruginosa*, which are well documented to cause eye infections. These may be transmitted by direct transfer by contaminated fingers or perhaps from aerosols. There are a number of issues here, notably aerosol minimization and personnel training. It is also prudent to recommend that safety spectacles are worn at all times in the microbiology laboratory.

4 Through the lungs. Again, steps to minimize aerosol generation and use and maintenance of biological safety cabinets are important here.

In the spirit of developing risk assessment based on previous experience it is worth noting the types of accidents which precede infection (Collins, 1993). These are spillage and splashes, needles and syringes, sharps and broken glass, and aspiration through pipettes. It is therefore essential to the whole risk assessment process that procedures are in place such that the potential for these accidents to occur is either removed or, at the very least, minimized.

The remaining steps of risk assessment centre largely on the maintenance of records and procedures to ensure the continued safety of the operator. These measures will include issues such as maintenance and testing of biological safety cabinets, environmental monitoring and personnel training before commencing a procedure and 'in-service'. The record of the assessment is clearly an important document and must be cross-referenced to Standard Operating Procedures (SOPs; see section 1.1.10). In the context of employee safety, it is important not to assume simply that the presence of SOPs guarantees safe practices. 'SOP fatigue' is well-recognized, and periodic checks should be carried out to ensure that procedures are followed. Finally, and perhaps most importantly, it must be recognized that risk assessment is a continuous process. It is not acceptable to continue a procedure over months or years without reviewing the risk assessment. A review process should be carried out at least once every five years, or more frequently if new information becomes available. This might include additional information on the pathogenic potential of the micro-organisms being used, or perhaps new innovations in laboratory safety equipment or changes in experimental methods.

### 1.4.3  *Risk ranking*

Clearly, not all risks are of the same importance and this is reflected in the rigour of the control measures to be adopted. In the case of handling micro-organisms this can be readily appreciated since the *likelihood* of exposure to organisms such as the harmless *Bacillus stearothermophilus* and the Category 2 pathogen *Pseudomonas aeruginosa* is the same, and yet the *consequence* of exposure to the latter is greater. In the event that it is not possible to deal with all risks equally, risk ranking is an approach to managing risks in the most important order of priority.

Risk level = Probability of occurrence × Severity of occurrence

Probability levels

    5    Certain or will occur at some time

    4    Probably or very likely to occur

    3    Possible or likely to occur

    2    Remote or unlikely to occur

    1    Improbable or very unlikely to occur

Severity levels

    5    Fatality

    4    Serious injury (hospitalization)

    3    Moderate injury ($>$3-day absence)

    2    Minor injury (first aid only)

    1    No injury

Combining the two risk elements gives the following risk levels:

| Risk levels 20–25 | are | CATASTROPHIC (Work prohibited) |
| Risk levels 15–19 | are | SERIOUS (Specific assessment) |
| Risk levels 10–14 | are | SIGNIFICANT (Specific or General assessment) |
| Risk levels 4–9 | are | MINOR (General assessment) |
| Risk Levels 1–3 | are | TRIVIAL (No written assessment) |

It is highly unlikely that risk levels in pharmaceutical QA procedures will extend beyond the 'minor' category, and so the various microbiological assays can be written into a single general assessment.

## 1.5 Genetically Modified Organisms (Contained Use) (Amendment) Regulations 1996

Although not relevant to micro-organisms currently used in pharmacopoeial tests, the reader should be aware that there is specific legislation concerned with all aspects of handling genetically modified micro-organisms (GMMs). Unlike COSHH 1994, a specific risk assessment procedure is described. The main requirements of the Contained Use Regulations provide for:

- human health and environmental risk assessment based on classification of the GMM into Group I or Group II (see below);
- records of risk assessments;
- establishment of a local Genetic Modification Safety Committee to advise on risk assessments;
- categorization of work into Type A and Type B operations (see below);
- advance notification to the HSE of an intention to use premises for the first time for activities involving genetic manipulation and, for some activities, consent from HSE before work can commence;
- notification to HSE of individual activities involving genetic manipulation and, for some activities, consent from HSE before work can commence; and
- standards of occupational and environmental safety and levels of containment.

Guidance on these Regulations is available (Health and Safety Executive, 1996). Briefly, a GMM is only classified into Group I if it satisfies all of the following:

'the recipient or parental micro-organism is unlikely to cause disease, neither the vector nor the recombinant DNA insert endows the GMM with a phenotype likely to cause disease and that the GMM is unlikely to cause disease to humans, animals or plants or cause adverse environmental effects'.

Failure on any part results in classification as a Group II GMM. The classification into the type of handling as Type A or Type B requires examination of tests of *purpose* and *scale*. A Type A operation typically involves small numbers of organisms which can easily be contained using good microbiological practice and which are readily inactivated by standard laboratory decontamination techniques. These activities might include teaching, research and development in either an academic or a commercial setting. It may also include development of a process for subsequent industrial or commercial exploitation. However, all operations which produce an industrial and commercial product are Type B operations, even though they may be carried out at small volume. In addition, large-scale research activities will also be Type B. The Regulations issue guidance on the measures required to meet the three Containment Levels, B2, B3 and B4, for Type B operations. Reviews by Hussey (1992), Atkinson *et al.* (1992) and Werner (1992) give a useful account of the issues relevant to handling recombinant micro-organisms in an industrial setting, however they do pre-date the current legislation.

## 1.6   Laboratory facility design

It should be apparent by now that risk assessment and safe microbial handling techniques are not an afterthought or 'add-on', simply to minimize risk in a less-than-ideal environment. Nowhere is this more important than in the design of the facility in which handling of micro-organisms is to be undertaken. A number of fundamental design elements are statutory requirements to operate a laboratory at Containment Level 2, whereas other design features, such as heating and ventilation, lighting, positioning of benches, etc. can make a significant contribution to safe practices.

### 1.6.1   *Work flow*

In the context of microbiological procedures within the pharmaceutical industry, consideration must then be given not only to the detail of containment laboratories, but also to 'clean room' suites for sterility testing and to the overall design of the facility in which the laboratories will operate. Good Manufacturing Practice (GMP) demands that the facility be separate from those areas involved in production. The logical work flow patterns implicit in GMP minimize the risk of cross-contamination of samples and ensure segregation of sterile material from contaminated waste and quarantined material from that passed for release. There are some useful guidance notes in the Annexes to Rules and Guidance for Pharmaceutical Manufacturers and Distributors 1997 (Medicines Control Agency, 1997). General considerations of laboratory design can be found in the texts by Collins (1993) and Liberman and Gordon (1989). The essential principles are:

- The laboratory must be designed to prevent access by unauthorized persons.
- Office areas should be separate from laboratories.
- There should be a clearly defined area for arrival and documentation of samples.
- Samples should follow a defined path through the laboratory: sample preparation/weighing, incubation, examination of cultures, autoclaving of contaminated waste and washing up.
- A hypothetical QC laboratory facility is shown in Figure 1.1.

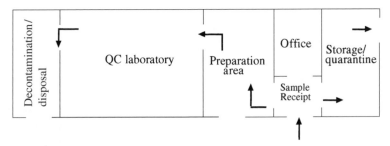

**Figure 1.1** Hypothetical Quality Control (QC) laboratory facility, showing restricted access to laboratory and work-flow design to minimize the risk of cross-contamination.

### 1.6.2 *Size and shape of rooms*

Sufficient space and carefully planned work flow are key elements in safe working practices. Adequate bench space is essential to guard against overcrowding, and 3 m per person is recommended. In terms of overall space, ACDP (Advisory Committee on Dangerous Pathogens, 1995) recommends a working space of at least 24 m$^3$ per person for laboratories operating a Containment Level 2 and higher. Where possible, preparation of sterile media, testing and decontamination should all be performed in discrete areas, and work flow patterns should be designed to ensure that cross-over cannot occur.

### 1.6.3 *Benches*

Each laboratory worker should have 3 m of bench space, though too much benching and inappropriate siting can be counterproductive to safe working practices. An approximately square laboratory with perimeter benching is preferable to extensive use of peninsular benching, which can make evacuation difficult in an emergency situation. As stated in the ACDP guidelines (Advisory Committee on Dangerous Pathogens, 1995), bench surfaces must be impervious to water and resistant to acids, alkalis and solvents. Raised edges are useful to contain spills, and all joints must be carefully sealed. Under-bench units should be movable for flexible use of space and cleaning.

### 1.6.4 *Floors, walls and ceilings*

Flooring should be impervious to liquids, anti-slip, seamless or have welded seams. Difficult-to-clean joints result in accumulation of residual dirt and contamination. They should be avoided by using sealed coving between floor and wall and wall and ceiling. Walls and ceilings should also be impervious to liquids, since splashes can arise from spillages. Epoxy resin- and polyurethane-type paints are suitable.

### 1.6.5 *Heating, lighting and ventilation*

A comfortable working environment is an important safety consideration, since the stress arising from conditions which are too hot or too cold is more likely to lead to

an accident. A temperature in the range 16–22°C is comfortable. Careful consideration should also be given to the positioning of ceiling lighting relative to operator
work area. Poor lighting and working in shadow will contribute to a deterioration in
safe working practices.

Ventilation is a critical issue and there is a fundamental difference between a
microbiological containment laboratory and a pharmaceutical 'clean room' suite. In
the latter, a positive pressure gradient, supplied by a high efficiency particulate air
(HEPA)-filter, exists between the different grades of 'clean room' (A/B, C and D), the
'step-over' white–grey–black area, changing room and the surrounding rooms at
normal atmospheric pressure. Here the objective is to achieve air quality levels
described in the Rules and Guidance for Pharmaceutical Manufacturers and
Distributors 1997 (Medicines Control Agency, 1997), and no attention is given to
the air escaping from the suite. It goes without saying that no micro-organisms,
irrespective of their containment categorization, would ever knowingly be taken into a
'clean room' suite. However, sterility testing procedures would be carried out in a
laminar air flow (LAF) cabinet within a dedicated unit, satisfying the requirement for a
Grade A local zone for high-risk operations within a Grade B background environment. Access to such an area would be restricted to all but authorized QC personnel,
who would change into clean area clothing in designated changing rooms. Design,
commissioning, operation, maintenance and monitoring of clean room suites are
beyond the scope of this chapter; however, they are described in BS 5295 (1989). An
explanatory supplement PD6609 (1996) gives additional information on the *in situ*
tests for HEPA filters in laminar flow cabinets, which should be performed at
6-monthly intervals.

For microbiological containment facilities, the philosophy of air handling is
completely opposite, and all attention is focused on the risk of pathogens in
contaminated air escaping from the laboratory. The ACDP (Advisory Committee
on Dangerous Pathogens, 1995) does not make mechanical ventilation a requirement
for operating a laboratory at Containment Level 2, recognizing that it is technically
difficult to maintain negative air pressure (inward flow of air) where workers enter and
leave the room frequently. However, when the ventilation system is designed to
contain airborne pathogens, then maintaining an inward flow of air is necessary only
when work is in progress. COSHH requires that a Containment Level 3 laboratory is
maintained at an air pressure negative to atmosphere.

## 1.7  Microbiological safety cabinets

Three classes of cabinet I, II and III are described in British Standard 5726: 1992.
These should not be confused with LAF cabinets which, for example, can be used to
create a Grade A aseptic filling and handling environment within a Grade B room in
a pharmaceutical manufacturing clean room facility, or alternately be used for sterility testing procedures (see above). Such cabinets protect only the product, offer no
protection to the operator, and should never be used for handling micro-organisms.
In a Class I cabinet air is drawn in through the front opening where the operator is
sitting, passes over the micro-organisms being handled, and is exhausted to atmosphere through a HEPA filter, complying with the requirements of BS 3928 (1969). A
Class II safety cabinet differs from that of a Class I cabinet in that 70–80% of the air
passing through the HEPA filter is re-circulated, with the remainder being exhausted

to atmosphere. Hence, the test material/micro-organism is bathed in HEPA-filtered air. For this reason, these cabinets are used extensively for tissue culture. One or both of these cabinets would be typically found in a Containment Level 2 facility and should be used where there is a risk of aerosols and airborne contamination. The Class III cabinet is totally enclosed, air is drawn in through a HEPA filter and exhausted by one, and usually two HEPA filters. The operator works through gloves which are an integral part of the cabinet. Such cabinets are used for handling Containment Level 4 organisms and would therefore be sited within a Containment Level 4 facility.

### 1.7.1 Siting and maintenance

Siting safety cabinets within the laboratory is an important issue, and critical work sites – away from exits and other 'laboratory traffic' – should be identified and used. Regular maintenance of safety cabinets, together with the associated documentation, must be an integral part of risk assessment, the SOP and good laboratory practice. Airflow testing procedures are described in BS 5726 Part 3. For a Class I cabinet the airflow through an unused HEPA filter must be between 0.7 and $1.0\,\mathrm{m\,s^{-1}}$ tested with a rotating vane anemometer. For a Class II cabinet the downward flow must be 0.25 to $0.5\,\mathrm{m\,s^{-1}}$ and the inward flow $0.4\,\mathrm{m\,s^{-1}}$. The maintenance schedules are described in BS 5726 Part 4. COSHH notes that cabinets should be tested at intervals not exceeding 14 months, and testing at least annually is appropriate.

### 1.7.2 Other laboratory equipment

All other pieces of laboratory equipment, such as incubators, pH meters and balances should have a log book which includes entries for mode of operation, calibration, date and operator use and schedule for routine maintenance. For equipment such as incubators, the temperature should be recorded at least once each day, and the reading and time of reading noted in the log book. The programme for calibration of equipment in the laboratory should be designed such that it can be traceable to national or international standards. For example, evidence of traceability could be through certificates issued by a national standards laboratory recognized or accredited by the National Measurement Accreditation Service (NAMAS). Details of NAMAS accreditation for microbiological testing are available (National Measurement Accreditation Service, 1992).

### 1.8 Sterilization, disinfection and decontamination

A cardinal rule, coined by Collins, is that *no infected material shall leave the laboratory*. This fundamental element to safe practice in the management of a microbiological QA laboratory should be designed into all SOPs and associated risk assessments. Again, texts by Collins (1993) and Collins *et al.* (1995) provide detailed guidance. Briefly, all infected material should be segregated and easily identified. The primary separation is into disposables and re-usables. Each should then be auto-claved or disinfected. Re-usables would then be washed and sterilized for return to the laboratory, whereas disinfected disposables would be incinerated and sent to a landfill site. Autoclaved disposables may not necessarily be incinerated prior to landfill. Colour-coding of laboratory waste should be used, and in the UK this is as follows:

- Yellow: for incineration
- Light blue or transparent with blue writing: for autoclaving
- Black: normal waste for local authority refuse collection
- White or clear plastic: soiled linen.

### 1.8.1  *Autoclaving*

Autoclaving is the method of choice for treatment of contaminated waste, although disinfection can be used in the laboratory to reduce the viable count prior to auto-claving. Laboratory autoclave performance is described in BS 2646 (1990, 1991, 1993) which includes Part I, Design and construction, safety and performance; Part II, Guide to planning and installation; Part III, Guide to safe use and operation; Part IV, Guide to maintenance; and Part V, Guide to testing function and perfor-mance. BS 3970 and BS EN 1422 cover sterilizers and disinfecting equipment for medical products. In most instances sterilization is achieved by exposure of the contaminated material to 121°C for 20 min. This should always be verified by insert-ing a calibrated thermocouple into the most dense part of the load. Timing should only begin once the temperature has been reached – this is often referred to in the literature as HTAT (holding time at temperature). Autoclave design and operation is beyond the scope of this chapter (but is discussed by Dewhurst and Hoxey, 2001); however, attention should be drawn to the type of container used for decontamina-tion. In the past, wire baskets have been used, as these offer the advantage of good air displacement properties; however, this advantage is outweighed by the lack of containment in the event of a spillage or breakage. Polypropylene buckets/skips are now widely used, although they should not be so tall that air displacement within the autoclave becomes a problem. Members of staff responsible for decontamination should be identified, and an SOP for decontamination should be understood and followed by all laboratory personnel. It is good practice to separate contaminated waste into disposables and re-usables at the point of generation, i.e. in the labora-tory, and load directly into autoclave skips – so minimizing contact by decontami-nation personnel. Sturdy autoclave skips, transported in stable trolleys satisfies the ACDP Containment Level 2 requirement that material for autoclaving should be transported to the autoclave in robust containers without spillage.

Commissioning and routine maintenance documentation in a logbook are of central importance to the safe operation of an autoclave and the safe handling of contaminated material. Routine testing is detailed in Health Technical Memorandum (HTM) 2010 (Department of Health, 1994). The same level of stringency should be applied to release of autoclaved contaminated waste as to release of an autoclaved pharmaceutical product. Process cycle records should compare favourably with the commissioning cycle record for a similar load before the material is released for incineration and/or landfill. Separate autoclaves for media sterilization and decontam-ination of waste material are desirable (BS 3970).

### 1.8.2 *Disinfection*

A variety of agents may be used in the laboratory in different circumstances depending on the micro-organisms being used and the level of other contaminating organic matter. There are a number of texts describing the various properties of these agents (Block, 1991; Scott and Gorman, 1992; Collins *et al.*, 1995). An important caveat here is that while such agents form an integral part of the specified disinfection procedures required for an ACDP Containment Level 2 facility and risk assessment for any given pharmaceutical microbiology QA procedure, many are in themselves hazardous substances, particularly in undiluted form. COSHH risk assessments should be made for these agents, and adequate personnel training is essential both to protect the laboratory worker and to ensure the efficacy of the agent. Having acknowledged that a variety of agents are available, hypochlorites are most widely used because of their broad-spectrum antimicrobial activity, which includes vegetative bacteria, spores and fungi. They are not compatible with cationic detergents and can corrode some metals. Discard jars should therefore be plastic and not metal. Activity is reduced by contaminating organic matter, and so the level of available chlorine should be varied depending on overall 'cleanliness'. For clean surfaces, such as benches, 1000 p.p.m. available chlorine is sufficient, whereas this should be increased to 2500 p.p.m. for pipette and discard jars. Industrial hypochlorites usually contain 100 000 p.p.m. available chlorine, whereas 'household' hypochlorites have between 10 000 and 50 000 p.p.m. available chlorine. Activity is lost rapidly on dilution, and so diluted preparations should be discarded after 24 h. Hypochlorites are irritant to skin, eyes and lungs, and hence particular care should be exercised when handling concentrates.

Alcohols are also used at a concentration between 70% and 80%. They are not effective against spores or fungi, only slowly active against vegetative bacteria, and are significantly inactivated by dilution; they are not, however, inactivated by organic matter.

Quaternary ammonium compounds (QAC) such as cetrimide are effective against vegetative bacteria and some fungi, but they are inactivated by a variety of materials, anionic compounds and soap. They are most active against Gram-positive bacteria and usually used at 1–2%. Alcoholic solutions are also used, combining the antimicrobial properties of both with the detergent-like properties of the QAC.

### 1.9 **Personnel and training**

Human resources are the key to safe working practices. It is essential for one senior member of the laboratory to hold an appropriate qualification, such as HNC, HTEC, BSc, AMILS and MIBiol, which all incorporate a major element of microbiology. All pharmaceutical QA laboratory staff should be trained and competent in handling micro-organisms and aseptic techniques, as well as having working knowledge of the individual test procedures and protocols. This latter aspect must be written into Standard Operating Procedures and will not be discussed further here.

Thorough instruction and practical training must be given to all staff handling micro-organisms. A number of personnel training issues are written into the ACDP requirements to operate a laboratory at Containment Level 1 or greater, including the cardinal rules of not eating, smoking or drinking in the laboratory, and no mouth pipetting. Although much has been written about safe handling of micro-organisms, the two critical components are hygiene and handling techniques.

### 1.9.1  *Hygiene*

All personnel must be made aware that contamination, particularly of the hands, can occur during normal routine procedures, even in the absence of an apparent spill. Protective clothing, in the form of a laboratory coat – preferably of the type which closes across the neck and has tight-fitting cuffs – should be worn at all times in the laboratory. These coats must not be worn outside the laboratory, and pegs should be provided for them near the exit. If they are contaminated they should be removed, placed in a suitably labelled bag and autoclaved, after which they can be laundered. Therefore, each worker should have sufficient coats to allow for this possibility, as well as the decontamination/laundry cycle which should occur at least twice weekly. The material, usually 100% cotton, must withstand autoclaving.

   Hair should be tied back. Jewellery and rings should not be worn, since they make thorough hand cleaning impossible. In order to reduce the risk of accidentally contaminating a product or test material, hands must always be washed prior to starting a procedure, and washed again on completion and upon leaving the laboratory. In the event of contamination occurring during a procedure then the hands must be washed immediately. If a worker has any cuts or grazes on the hands, these should only be of a superficial or minor nature and be covered with a waterproof dressing for work to be allowed. The issue of wearing disposable gloves, usually latex, has been controversial and gloves are not necessary for handling the ACDP category 1 and 2 organisms used in pharmaceutical QA procedures. The principal criticism of glove wearing centres on the notion that operators can forget that, while they are protected, they can very easily spread contamination around the laboratory to such items as pipetting devices, refrigerator and freezer door handles and the telephone. Since some of the most common means of laboratory-acquired infection involve contact with the face (eye infections and ingestion), all personnel must be trained and instructed not to touch the face or hair while handling micro-organisms. Obviously, sterility testing in a clean room requires staff to change into specialized clothing, following the SOP for changing.

### 1.9.2  *Handling techniques*

Laboratory personnel should be confident with aseptic techniques, that is performing relevant manipulations without introducing contaminants, before proceeding to handle micro-organisms of any type. Training in routine liquid handling and filtration techniques can be undertaken with a variety of pipetting aids used in the laboratory and sterile growth medium such as tryptone soya broth. Clearly, any contamination will be readily apparent after suitable incubation. While describing and demonstrating the use of pipetting aids and 'tools-of-the-trade' such as bacteriological loops it is important to introduce the concept, and awareness of the danger, of aerosol generation. For example, graduated pipettes should not be of the 'blow-out' type, and personnel should be trained to release liquid from any type of pipette carefully down the side of the vessel. Liquid can be released under the surface of another liquid, for example when performing a serial dilution in buffer for a total viable count (TVC); however, care should be taken not to introduce bubbles, since many automatic pipettes are of the 'second-push' type to deliver the correct volume. Any procedure which carries an unavoidable risk of aerosol generation must be performed in a biological safety cabinet. Much has also been written about the aerosol risk from bacteriological loops, including loop diameter and distance from the handle. If a loop carrying micro-organisms is introduced directly into a sterilizing flame (Bunsen burner), then there is a significant risk of aerosol generation. Wire loops should be sterilized by first heating the platinum wire near the handle and then slowly drawing the loop into the flame until it is red hot. Once aseptic technique has been established, a training programme using a category 1 organism, such as a laboratory strain of *Escherichia coli*, should be undertaken if the worker has no previous microbial handling experience. Such a programme might comprise inoculation and plating techniques, including a TVC.

Beyond this training in safe bench practices, staff must also receive training in the following:

1  Behaviour and conduct in the laboratory.

2  First aid.

3  COSHH and the handling of chemicals and cultures.

4  Laboratory housekeeping, disinfection procedures and waste management.

5  Use of equipment.

6  Media preparation.

7  Sample documentation and record-keeping.

Training must not be regarded as a 'one-off' event by either the employer or the laboratory personnel. Good staff motivation is essential to prevent lapses in good practice and SOP fatigue. There should be good communication with other departments, particularly production so that staff can appreciate how they contribute to the overall QA programme of the organization. Moreover, it is essential that staff appreciate the importance of documentation, following procedures and correct product identification at all times. Employers must have a documented training programme, which should define levels of competence to be attained before staff are permitted to work without direct supervision. Records of training undertaken by personnel should also be kept so that staff development needs can be identified, as well as for suitably trained and qualified personnel to substitute for a colleague if necessary. Laboratories should also operate an internal quality control scheme to monitor the performance of individual microbiologists. This might take the form of cross-checks between analysts, analysis of spiked samples, and duplicate analysis of a single sample.

## 1.10   Documentation

### 1.10.1   *Standard Operating Procedures*

References have already been made to the importance of record-keeping and documentation, in particular SOPs. Under the demands of QA within Good Manufacturing Practice it is essential that accurate documentary evidence is taken throughout the manufacturing and quality control processes to ensure the product meets the required specification and enable fault tracking in the event of product failure. Records should be made with sufficient rigour to ensure that they withstand self audit and external inspection, for example by the Medicines Inspectorate. Central to any manufacturing process is the SOP, since it is adherence to these procedures which is the basis of QA. Within the pharmaceutical microbiology QA laboratory, there should be SOPs for all of the following:

- all laboratory equipment, including operation, calibration and the maintenance schedule;
- training of all staff;
- reception of visitors;
- laboratory housekeeping and hygiene, including cleaning and environmental monitoring;
- sample receipt, record-keeping, handling and disposal;
- reagents and microbiological media – preparation, shelf-life and quality control;
- methods;
- waste management;
- handling of cultures and dealing with spillages;
- accidents. Specifically here an accident report form should be completed and logbook kept of all incidents. Each accident form should be brought to the attention of the Safety Team for review and possible amendment of hazardous procedures;
- documentation and release of batches from quarantine.

Clearly, documentation of some of these procedures, which was formerly good practice, is now a requirement under COSHH.

SOPs should be designed such that a suitably trained technician can complete the task safely and correctly. A well-designed SOP should include the following elements:

1  The title should be a clear description of the procedure to be undertaken. For example, a SOP for a piece of equipment should include the specific model number, and it is implicit in this that it refers only to operation of that particular model.

2  The staff responsible for ensuring that the SOP is followed should be clearly identified. The level of staff training required for the SOP to be undertaken should also be stated.

3  The purpose of the SOP should be readily apparent.

4  A clear step-by-step guide to use in language that is not open to misinterpretation. It may be useful to 'test' the SOP with a number of staff who are unfamiliar with the procedure to confirm this point.

5  Where calibration is important, the procedure should be described in full and include the acceptable tolerance limits. The criteria for failure of calibration and normal operation should be clearly stated.

6  In the event of failure, the procedure for rectification, including the level of staff authorization required for remediation, must be clearly defined.

7  The appropriate risk assessment, required by COSHH, must also be included with the SOP.

### 1.10.2  Quality Systems

Safety issues are inextricably linked to considerations of quality. Increasing emphasis has been placed on the importance of setting up and operating a Quality Management System (QMS) in both the medical device and pharmaceutical industries in recent years.

The laboratory should operate a Quality System appropriate to the type, range and volume of work performed. The elements of the System should be maintained in a Quality Manual by a responsible member of the laboratory personnel. This person will be nominated by the laboratory management team and have direct access to top management. The Quality Manual shall contain at least the following elements:

1  A quality policy statement.

2  The organization and management structure of the laboratory.

3  The operational and functional activities relevant to quality, so that personnel appreciate the extent and limits of their responsibility.

4  Reference to quality assurance procedures specific for each test (e.g. SOPs)

5  Procedures for achieving traceability of measurements.

6  Procedures for dealing with complaints.

7  Procedures for audit and review.

### 1.11  Putting theory and legislation into good laboratory practice

Having described the legislation and some of the theoretical issues concerned with safe handling of micro-organisms, the challenge facing the QA microbiologist in the pharmaceutical industry is to channel these into day-to-day good laboratory practice. Appended below is a check-list which should be followed:

1  At the design and building stage, a team including a microbiologist should be assembled to ensure compliance with building regulations, health and safety legislation and the requirements to operate a laboratory at Containment Level 2.

2  All major pieces of equipment, such as LAF cabinets, biological safety cabinets, autoclaves, should be appropriately commissioned, subsequently monitored, and all documentation maintained.

3  Standard Operating Procedures must be written for all pieces of equipment and procedures. These must include not only specific test procedures, but also preparative procedures such as media preparation and, most importantly, 'clean-up' procedures such as disinfection and decontamination. The procedures for documentation should be such that there is communication with other

departments, particularly for follow-up of out-of-specification results, and facilitate trend spotting.

4   Local safety rules should be drafted, including the names and contact addresses of persons with overall responsibility for safety. In an industrial setting, these will have been written at the company level, but it may be necessary to make additional comments, for example on routes of evacuation from clean room facilities.

5   Notification to HSE of intention to propagate Category 2 organisms.

6   If genetically modified micro-organisms are to be used, then a Genetic Modification Safety Committee will have to be established and a Biological Safety Officer appointed. Notification of intention to commence activities will have to be made to the Health and Safety Executive.

7   Once appointed, staff should receive training in the procedures to be undertaken and periodic 'in-service' training.

8   A process of self-audit should be commenced to ensure that samples for testing and analysis can be tracked from the moment that they are received by the laboratory to the ultimate fate of the batch. Implicit in this is documentation of the release of a batch passed fit for use, or the destruction of a failed batch.

## 1.12   Summary

This chapter has described the current legislation and good practices surrounding the safe handling of micro-organisms. Inevitably, not every facet can be covered; however, if appropriate attention is given to risk assessment, then most of the hazards associated with handling micro-organisms can be designed out of pharmaceutical QA laboratory procedures. Once the QA laboratory is operational, then all procedures should be documented and all personnel trained. A useful maxim is *if it is not documented, or personnel are not trained, don't do it!*

## References

Advisory Committee on Dangerous Pathogens (1995) *Categorization of biological agents according to hazard and categories of containment.*, 4th edn. HSE Books, London.

ATKINSON, T., CAPEL, B.J. and SHERWOOD, R. (1992) Recombinant DNA techniques in production. In COLLINS, C.H. and BEALE, A.J. (eds), *Safety in Industrial Microbiology and Biotechnology*. Butterworth-Heinemann, Oxford, pp. 161–75.

BLOCK, S.S. (1991) *Disinfection, Sterilization and Preservation*, 4th edn. Lea & Febiger, Philadelphia.

British Standard 2646 (1990, 1991 and 1993) *Autoclaves for sterilization in laboratories*. British Standards Institution, London.

British Standard 3928 (1969) *Method for sodium flame test for air filters (other than for supply to i.c. engines and compressors)*. British Standards Institution, London.

British Standard BS 3970 (1990) *Steam sterilizers*. British Standards Institution, London.

British Standard 5295 (1989) *Environmental cleanliness in enclosed spaces*. British Standards Institution, London.

British Standard 5726 (1992) *Microbiological safety cabinets*. British Standards Institution, London.

British Standard EN 1422 (1998) *Steam sterilizers for medical purposes – Ethylene oxide sterilizers – requirements and test methods*. British Standards Institution, London.

British Standard PD 6609 (1996) *In situ testing of HEPA filtrations. An explanatory supplement to BS 5295: Part 1: 1989*. British Standards Institution, London.

COLLINS, C.H. (1993) *Laboratory-Acquired Infections*. Butterworth-Heinemann, Oxford.

COLLINS, C.H., LYNE, P.M. and GRANGE, J.M. (1995) *Collins and Lyne's Microbiological Method*, 7th edn. Butterworth-Heinemann, Oxford.

Department of Health (1994) *Health Technical Memorandum 2010*. HMSO, London.

DEWHURST, E. and HOXEY, E.V. (2001) Sterilization methods. In: DENYER, S. and BAIRD, R. (eds), *Guide to Microbiological Control in Pharmaceuticals*, 2nd. edn, Ellis Horwood, London, in press.

Health and Safety Executive (1996) *A guide to the Genetically Modified Organisms (Contained Use) Regulations 1992, as amended in 1996*. HSE Books, London.

HUSSEY, C. (1992) Recombinant plasmids. In: COLLINS, C.H. and BEALE, A.J. (eds), *Safety in Industrial Microbiology and Biotechnology*. Butterworth-Heinemann, Oxford, pp. 93–152.

LIBERMAN, D.F. and GORDON, J.G. (1989) *Biohazards Management Handbook*. Marcel Dekker, New York.

Medicines Control Agency (1997) *Rules and Guidance for Pharmaceutical Manufacturers and Distributors*. HMSO, London.

National Measurement Accreditation Service (1992) *NIS 31. Accreditation for Microbiological Testing*. NAMAS Executive, London.

SCOTT, E.M. and GORMAN, S.P. (1998) Chemical disinfectants, antiseptics and preservatives. In: HUGO, W.B. and RUSSELL, A.D. (eds), *Pharmaceutical Microbiology*, 6th edn. Blackwell Scientific Publications, Oxford, pp. 201–28.

WERNER, R.G. (1992) Containment in the development and manufacture of recombinant DNA-derived products. In: COLLINS, C.H. and BEALE, A.J. (eds), *Safety in Industrial Microbiology and Biotechnology*. Butterworth-Heinemann, Oxford, pp. 201–28.

*United States Pharmacopoeia XXIII* (1995) United States Pharmacopoeial Convention, Rockville, MD.

**2**

# Culture Media Used in Pharmaceutical Microbiology

ROSAMUND M. BAIRD

*Department of Pharmacy and Pharmacology, University of Bath, Claverton Down, Bath BA2 7AY, United Kingdom*

## 2.1  Introduction

Microbiologists working today have considerable choice in the range and type of culture media available to them. This has largely arisen from the expansion of microbiology from medicine in its early days to agriculture, food manufacturing, water production and pharmaceutical applications, with each discipline having its particular and individual requirements for culture media. The design and subsequent development of culture media have largely reflected these requirements, but have also been influenced by the introduction of selective and differential indicator media, containing an array of inhibitory substances in various quantities and combinations. In addition, culture media, initially developed for a particular requirement, have sometimes later been successfully adopted for use in other disciplines of microbiology. In the case of pharmaceutical microbiology, however, a rather more conservative approach has been taken over the introduction of new media; the range of media used has remained largely consistent over the years, undoubtedly reflecting in part pharmacopoeial recommendations.

Culture media are used for the isolation, enumeration and identification of micro-organisms (see Chapters 4 and 5).

## 2.2  Design of media

There is no universal culture medium which will support the growth of all micro-organisms. Equally, many environments – including seemingly hostile ones such as pharmaceutical products – will provide sufficient nutrients to enable selected micro-organisms to grow. Even distilled water contains sufficient organic matter to support the growth of *Burkholderia cepacia* (previously known as *Pseudomonas cepacia*) to high numbers when held at room temperature over a number of days (Eisman *et al.*, 1949). In order to survive, micro-organisms must derive all their cellular constituents from their immediate environment. This may range from a simple salts solution to a complex and lengthy list of formulation components.

In the main, however, a chemically defined culture medium may contain the following:

- an energy source in the form of carbohydrates and commonly glucose;
- amino-nitrogen nutrient source in the form of peptones, other protein hydrolysates or extracts that will provide a source of carbon, nitrogen and energy;
- growth-promoting factors, such as blood, serum, vitamins, are required for the growth of fastidious organisms;
- metals and mineral salts, such as Ca, Mg, $Fe^{+++}$, $Mn^{+++}$, trace metals, phosphates and sulphates. These inorganic elements are added at macro or micro levels to enhance microbial growth or population yield;
- buffer salts, in the form of acetates, citrates, phosphates, will maintain pH stability where fermentation occurs;
- indicator dyes, such as bromocresol purple, neutral red or phenol red will indicate a shift in media pH, following the growth of micro-organisms. The concentration of these agents is critical, since they may be toxic to some organisms;
- selective agents, i.e. various antibiotics, chemicals (such as bismuth sulphite, cetrimide, lithium chloride, potassium tellurite, sodium selenite, tetrathionate), inhibitory dyes (such as acriflavine, crystal violet, brilliant green) are added to suppress the growth of unwanted organisms, and concurrently promote the growth of desired organisms. The concentration of these is also critical;
- gelling agent, i.e. agar, is added to give a solid inert gel on which organisms can grow in separate colonies. In the case of a pour plate inoculum method, the molten agar retains dispersed organisms within its matrix until the gel sets.

Growth of micro-organisms will also be determined by their requirements for oxygen – aerobic organisms such as *Pseudomonas* growing in a positive range of the redox potential, anaerobic organisms such as clostridia growing only at a low redox potential, and facultative anaerobes such as *Enterobacteriaceae*, growing over a range from +150 mV to −600 mV. Temperature and pH will also have a profound effect on the microbial growth characteristics.

A detailed discussion of the function of media constituents is outside the scope of this chapter, but several excellent reviews have been published (Prescott *et al.*, 1993; Bridson, 1994; Mossel *et al.*, 1995).

### 2.2.1 *Specialized media for neutralizing antimicrobial agents*

Any antimicrobial agent present in products must first be inactivated before testing can proceed. As shown in Table 2.1, this can be achieved in four ways: (i) the cells can be physically separated from the product by filtration, if this is appropriate; (ii) in the case of antimicrobial agents with a high dilution coefficient (e.g. alcohols, phenols), the activity may be diluted out; (iii) when antibiotics are present, enzyme inactivation may be employed, as in the case of $\beta$-lactamase used to inactivate cephalosporins; and (iv) for other antimicrobial agents a specific or non-specific inactivating agent may be added. Lubrol W (4%) and Polysorbate (Tween) 80 (3%) are commonly used when no specific neutralizing agent is available. Media

**Table 2.1**  Methods of inactivating commonly used antimicrobial compounds

| Antimicrobial compound | Method/addition of |
|---|---|
| Alcohols | Dilution (polysorbate (Tween) 80) |
| β-Lactam antibiotics | β-Lactamase from *Bacillus cereus* |
| Antibiotics, other | Membrane filtration |
| Bronopol | Cysteine hydrochloride/glutathione |
| Chlorhexidine | Lecithin + polysorbate 80 or Lubrol W |
| Cresols | Dilution + polysorbate 80 |
| Halogens | Sodium thiosulphate |
| Hexachlorophane | Polysorbate 80 |
| Mercurials | Sodium thioglycollate or cysteine |
| Parabens | Dilution + polysorbate 80 |
| Phenolics | Dilution + polysorbate 80 |
| Quaternary ammonium compounds | Dilution + lecithin + polysorbate 80 |
| Sorbic acid, benzoic acid | Dilution + polysorbate 80 |
| Sulphonamides | *p*-Aminobenzoic acid |

containing neutralizing agents are commercially available, e.g. D/E Neutralizing Agar and Letheen Agar (Difco, BBL Products). The former was developed specifically to neutralize residual antimicrobial agents, and has been used in disinfectant evaluation, environmental sampling and in product testing (Curry *et al.*, 1993; Dey and Engley, 1994).

### 2.2.2  *Specialized media for recovering injured cells*

Micro-organisms recovered from manufactured items may well have been exposed to the injurious effects of processing; these may include a variety of physical treatments, such as freezing, heating, dehydration or the effects of extremes of pH, or osmotic pressure, or chemical treatments, such as the addition of antimicrobial agents or preservatives. Microbial isolates from pharmaceutical preparations as well as from medical devices are highly likely to have encountered such processing in the course of their manufacture, and frequently therefore contain a proportion of so-called 'stressed' populations. In the past, the significance of these stressed populations has been debated; nowadays, however, their importance is well recognized (Russell, 1991).

The sites of cellular damage in vegetative cells and spores which have been exposed to the injurious effects of processing have been reviewed (Hurst, 1977). Such cells are characterized by:

- an increased sensitivity to surface active compounds, salts and toxic chemicals, antibiotics, dyes, acids and low pH;
- a loss of cellular materials;
- an extended lag phase;
- an ability to repair injury if provided with suitable conditions; and
- an inability to multiply, until repaired (Ray, 1979).

In other words, these populations may fail to recover in unfavourable circumstances, i.e. if the correct recovery conditions are not provided, leading perhaps to a gross under-estimate of contamination levels in the product concerned. Furthermore, if damaged cells are subcultured directly onto selective media, there may be no recovery at all. An additional feature of stressed cells is that they are also rendered more sensitive to a secondary stress. For example, cells which have been exposed to the antimicrobial activity of a preservative are likely to be more sensitive to cold shock if placed in a diluent which has been taken directly from the fridge. The resuscitation process therefore must provide a suitable recovery medium to enable cell repair to occur during a sufficient time at the correct temperature (Dodd *et al.*, 1997; Denyer and Stewart, 1998).

There is no universal resuscitation medium. In general, the use of complex – rather than simple – media is recommended as damaged organisms may be able to utilize a wider choice of substrates for alternate metabolic pathways in their recovery. In food microbiology an extensive number of recovery agents have been added to culture media and evaluated for their usefulness (Mossel and Corry, 1977); by contrast, in pharmaceutical microbiology the subject has barely merited discussion until comparatively recently. In the current *British Pharmacopoeia* (BP) (1999), neither the sterility test nor the preservative efficacy test include a reference to the likely presence of damaged organisms and how they might be recovered from test samples.

Recovery agents can be divided into various categories:

1 general recovery agents, e.g. whole blood, egg yolk, starch, oxygen radical scavengers such as charcoal;

2 specific neutralizing agents which can inactivate residual chemicals (see section 2.2.1 above);

3 catalase (added as the enzyme or as whole blood) to remove peroxides;

4 lysozyme permitting the germination of apparently dead spores;

5 pyruvate and other citric acid cycle intermediates, where membrane damage and subsequent leakage of tricarboxylic acid (TCA) intermediates has occurred;

6 magnesium used to stabilize RNA and cell membranes.

Both liquid and solid medium repair methods can be used. In liquid medium, phosphate-buffered peptone water or tryptone soya broth can be used; the recovery conditions (time and temperature of incubation) would depend on the isolate concerned. The BP (1999) recommends that the sample should be pre-treated by incubating in lactose broth at 35–37°C for a period usually between 2 and 5 h when examining for the presence of Enterobacteriaceae and other Gram-negative bacteria in non-sterile products. Similar recovery conditions are used in the examination of food samples for microbial contamination.

## 2.3 Good laboratory practice in culture media preparation

### 2.3.1 *Media preparation*

Culture media may originate from one of three sources: laboratory-prepared from component raw materials, commercially manufactured and presented in either a dehydrated, or ready-to-use form. The first of these is now rarely encountered in the pharmaceutical industry, and most microbiology quality assurance laboratories will employ dehydrated or ready-to-use products. However, in some instances where unusual formulations are required, there may be no alternative to the use of laboratory-prepared media. Pharmaceutical microbiologists use a narrow range of media in the main, all of which are available from a commercial source. The decision of whether to use a dehydrated medium concentrate or a short-life, ready-poured culture medium will be influenced by considerations of economy, space and available staff time for both production and quality control. Whatever the source, the microbiologist's expectations of his or her culture media remain the same: a freshly prepared quality product in all instances.

### 2.3.2  *Storage of incoming raw materials*

All incoming raw materials should be purchased from an approved supplier and dated on receipt. Particular care should be given to the sourcing of media used in the production of pharmaceutical materials with respect to the bovine spongiform encephalopathy status of herds providing beef extracts (Garland, 1999). The suitability and security of packaging should be checked. Raw materials should be stored as advised by the manufacturer; humid environments (in the vicinity of steam pipes) and those with fluctuating temperatures are clearly unsuitable.

### 2.3.3  *Water*

Water quality can markedly influence the performance of culture media. Fresh, good quality water must be readily available and provided by distillation, deionization or reverse osmosis. Tap water is unsuitable owing to its inherent impurities (traces of magnesium, calcium and other metals, as well as chlorine and fluorine), all of which may affect the characteristics of selective culture media. Water quality should be regularly monitored.

   Conductivity measurements should be made daily and should be within the manufacturer's specified low range. Higher readings suggest that equipment maintenance may have been neglected. The pH of water is not usually checked, unless there is a problem with the pH of prepared media. As with pharmaceutical grade water, water quality can only deteriorate on storage, and therefore ideally supplies should be produced and used as required. If water must be stored, clear glass containers with lids are preferable to polythene containers as the pH may become acid on storage.

### 2.3.4  *Preparation of culture media*

Good laboratory practice requires that culture media are made to pharmaceutical and not 'cook book' standards. Media should be made according to master formulae and written procedures. Documentation should be completed for each batch of media made; a batch number and expiry date should also be allocated. Manufacturer's instructions should be closely followed.

   When weighing out powders, a suitable face mask should be worn by the operator. Excessive dust can be avoided by weighing out each ingredient into a suitable weighing boat. Depending on the quantity involved, a top pan balance (accuracy $\pm 0.01$ g) or analytical balance for smaller weights will be required. Where very small quantities are involved, it may be necessary to prepare a concentrated stock solution and then dilute this to the required strength. When dissolving powders in water, the aqueous phase (1/3 volume) should be added first, followed slowly by the powder, dispersed by gentle swirling, and finally the remaining liquid can be used to wash down any adhering powder.

Balances should be cleaned thoroughly after use; they also require weekly calibration and regular maintenance. Appropriate records should be kept, as discussed below. Equipment used for dispensing culture media should be thoroughly cleaned after use.

### 2.3.5 *Measurement of pH*

The properties of a given culture medium are pH-dependent. Incorrect pH may result not only in physical changes such as precipitation of components or soft gelling of agar, but also significant chemical changes, such as a loss or change in the indicator or selective system. Incorrect pH may also affect the recovery of stressed cells and influence cell growth.

Since pH is temperature-dependent, measurements are best taken at a standardized temperature, e.g. 25°C using a suitable instrument (i.e. measuring to 0.1 pH units). pH meters should be calibrated before use using standardized buffer solutions (e.g. pH 4, 7 and 10) and these should be renewed at least weekly. As before, planned preventative maintenance is required, and records should be maintained. The pH of laboratory-prepared media should be checked and adjusted if necessary before dispensing for sterilization. Changes in pH may occur during sterilization; allowance for this can be made on the basis of experience.

In contrast, commercially available dehydrated media usually require no pH adjustment if properly prepared. The pH of the final sterilized product can be measured on a single unit (plate or bottle), but this must be discarded after use. pH measurements should be recorded in the batch records.

### 2.3.6 *Sterilization*

The sterilization of culture media is a critical control point in assuring their quality: the media must be sufficiently processed to ensure sterility, but any over-processing may affect their nutritive properties and result in the accumulation of toxic residues. Thus, as in pharmaceutical production, individual sterilization cycles should be properly validated with thermocouples to ensure that all containers in the load achieve the required temperature. Furthermore, for each batch processed the accompanying temperature time chart recording should indicate that the correct conditions in the load have been achieved. These charts should be retained with the accompanying batch documentation.

### *Autoclaves*

Loading patterns in autoclaves should be considered to allow free circulation of steam and adequate steam penetration. Similar volumes should also be processed at the same time whenever possible. Correct segregation of sterilized and unsterilized media should be ensured through a streamlined work-flow process. Autoclave tape may also be used as a supplementary aid, but should not be regarded as hard evidence that a satisfactory sterilization cycle has been achieved.

Guidance on the design, installation, operation, maintenance and process monitoring of autoclaves is available in BS 2646 (1988, 1990), and the importance of following these recommendations cannot be overemphasized. As before, all records should be retained.

### Agar preparators

The sterilization of large volumes (i.e. in excess of 1 litre) of culture media in an autoclave can result in a nutritionally inferior product if an extended heat transfer occurs during heating up and cooling down. In contrast, agar preparators provide greater control of the heating process with rapid heating and cooling cycles, resulting in a good quality product. Here, liquid medium is introduced directly into the jacketed heating chamber, continuously mixed throughout the sterilization cycle, cooled rapidly and then pumped directly into containers. After use, the bowl must be thoroughly cleaned.

### Membrane filtration

Membrane filtration is used to sterilize heat-sensitive media and components. Filters with a pore size of 0.45 or 0.22 µm may be used, supplied in a ready-to-use disposable filtration unit or as a re-usable and sterilizable filter unit with a separate filter.

### 2.3.7  *Plate pouring*

Depending on the chosen method of isolating and counting microbial colonies, the product sample will either be spread as a dilution onto the surface of a pre-poured set and dried agar plate – a technique known as surface inoculation – or alternatively the sample can be mixed with molten agar, then poured into a Petri dish and allowed to set – known as the pour plate technique. In both instances the agar should be carefully tempered to $46 \pm 1°$C. Above this temperature, there may be damage to microbial isolates or heat-labile supplements, or excessive condensation may occur; below this temperature, the gel may begin to set before plates have been poured. Care should be taken to avoid the introduction of air bubbles by gently swirling the contents to ensure thorough mixing of ingredients and possible supplements. When pouring plates by hand, an aseptic technique should be used and ideally within a laminar flow cabinet. A 15–20 ml volume is required in each 90-mm Petri dish, in order to minimize batch-to-batch variations in microbial performance. Smaller volumes may result in the medium drying out. Automated plate pouring systems may be used where there is a high demand for plates of a limited range of media types. However, before embarking on such a system, the microbiologist should consider the plate throughput per hour, the destacking and stacking mechanisms for loading plates, the likelihood of blockage and ease of clearing, the acceptability of Petri dishes from alternative suppliers, and finally the proposed savings in staff time offset against the capital cost of the system.

## 2.4 Prepared media storage

All media should be used within their given shelf-lives. The shelf-life of media is determined by a number of factors: the type of media and the container, as well as the storage conditions. Freshly prepared media undoubtedly give the best performance, but their use may not always be practical. Where appropriate, manufacturer's instructions should always be followed. Bottled liquid media in air-tight containers may be stored for 2–4 months in the dark at 2–8°C; at room temperature the shelf-life will be much shorter. In contrast, the shelf-life of prepared plates will be determined by the stability of added supplements. Agar plates begin to deteriorate from the moment of preparation; however, correct packaging combined with low-temperature storage will slow the rate of deterioration, giving an acceptable storage life in the laboratory. Stored upside down in sealed polythene bags in the dark at between 2–8°C, these plates can safely be kept for several days; at room temperature their life is limited to 2 days. If the shelf-life has to be extended, the nutritive and selective properties of the media should be confirmed. Loss of moisture from prepared plates is another consideration, perhaps resulting in concentration changes in key components. Shrinkage in the depth of plate agar from 5 mm to 4 mm during storage will at the same time ensure that component raw materials, including any antimicrobial agents, become 20% more concentrated.

Before using stored plates, the agar should be checked for the absence of contaminants. Plates showing shrinkage or wrinkling of agar should be discarded; wet plates should be dried until visible moisture has disappeared.

Stored, solid media may be re-melted for use by placing in a boiling water bath or free steaming in an autoclave at atmospheric pressure. Media must not be re-melted by the direct application of heat. The minimum exposure to heat should be used to avoid compromising nutritive and selective properties. Once molten, the media should be allowed to cool to $46 \pm 1$°C. When re-melting medium which has previously been held in chilled storage, it should first be tempered to avoid cracking of containers on exposure to heat.

## 2.5 Plate drying

In drying plates, a balance must be struck between removing visible surface moisture from the agar and at the same time avoiding over-dry plates; the latter may result not only in increasing concentrations of nutrients and selective agents, thereby impairing microbial growth, but also a reduction in the amount of available water, thereby decreasing colony size. Plates can be dried open, preferably in a laminar flow cabinet or in an incubator, for short periods of time; alternatively, vented plates can be dried closed at ambient temperature in the dark or in an incubator for longer periods. The practice of incubating plates overnight at 35–37°C to confirm sterility is known to compromise the growth performance of media and should therefore be discouraged. When using a laminar flow cabinet, it should first be prepared according to the manufacturer's instructions, before the plates are laid out with the agar surface facing up and the lid facing down. When dried open in an incubator, plates and lids should be separated and both placed facing downwards to avoid contamination settling onto the exposed surfaces. Once dried, plates should be used promptly. The practice of leaving dried plates on the bench at room temperature for several hours may provide an opportunity for growth of any airborne contaminants originating from the incubator into micro-colonies. These may subsequently be dispersed when a glass spreader is used to distribute the sample over the agar surface.

## 2.6   Use of media

Before use, media must be at the correct temperature if thermal or cold shock is to be avoided. Hot molten agar must be tempered to $46 \pm 1°C$ before use if heat inactivation of microbial isolates is to be avoided with ensuing low microbial counts. Furthermore, the activity of heat-sensitive supplements may be affected, resulting in denaturation and an associated reduction in the selective/differential properties of the agent. As mentioned previously, excessive condensation may also result, requiring extended drying of plates. If tempering in a water bath, the water level should be at the same height as that of the medium to avoid uneven cooling, but not so high that it enters the container. The water should be changed regularly to avoid the build-up of microbial contamination. Once the required temperature has been reached, the media should be used as soon as possible, and certainly within a 4-h period. Likewise, chilled media, including diluents, should be allowed to equilibrate to room temperature before use.

## 2.7   Records

Records should be maintained of all media received and used in the laboratory. These should include the name of the medium, batch number, date received, storage temperature, date used, quality tests carried out and results, performance of the medium in use, number of contaminated units, and the number of other defects. The nature of any complaint and the manufacturer's response should also be noted in the laboratory records.

  For media manufactured in the laboratory, the batch record should detail the date of manufacture, the quantity and batch number of each raw material used, the total quantity of medium prepared, the number of units filled and wasted, the batch number and expiry date allocated, and the pH and appearance recorded; the accompanying sterilization charts should be retained with the batch records. The results of all quality control tests should also be logged.

## 2.8   Quality control of culture media

### 2.8.1   *Manufacturer's responsibility*

Responsibility for quality rests primarily with the manufacturer of culture media. Acknowledging this, reputable manufacturers have in the main today embraced the principles of Total Quality Management (TQM), which provides the framework for a rigorously defined quality procedure that controls not only raw materials sourced world-wide, all manufacturing operations, as well as laboratory analysis and control, but also the follow-up of complaints and defects.

With this framework in place, the laboratory manager can then devote his or her attention to the internal quality control tests to be carried out on each batch of culture media, involving a range of physical, chemical and microbiological tests. Physical tests should confirm that the powder meets its specification in terms of colour, odour, particle size, homogeneity, flow characteristics and moisture content. Chemical tests would include the colour, clarity and solubility and pH of the product when reconstituted with deionized water, sterilized and cooled, as required. Media containing agar must also meet the required specification for gel strength. Various other tests may also be carried out which relate specifically to the medium concerned.

With regard to microbiological tests, there are to date no specified or mandatory growth tests by which the performance characteristics of culture media are measured in pharmaceutical microbiology. Most manufacturers do, however, provide a list of suitable test organisms for the end-user to monitor performance of the media concerned. However, the Pharmacopoeia of Culture Media (Corry *et al.*, 1995) lists the minimum test organisms recommended to be used for monitoring 74 of the more common media used in food microbiology. A list of supplementary strains of test organisms for more extensive testing is also given when assessing new batches of media or when investigating problems. Similarly, suitable test organisms for monitoring 250 media in medical microbiology have also been listed (MacFaddin, 1985). Suitable test methods for use in the microbiological control of both liquid and solid culture media are also described in the Pharmacopoeia of Culture Media. Test criteria are also given, based on the definition of satisfactory growth of a test organism on a non-selective medium which should be within $0.7 \log_{10}$ of the count of the same strain on the reference medium. In the case of selective media, the recovery of unwanted test strains should be at least 5 log units below the recovery on the reference medium. The comparison of test versus reference medium is an essential part of any quality control testing scheme, thereby ensuring that any variations in inocula, viability or phenotypic characteristics are controlled for each test. Pharmaceutical microbiologists have yet to turn their attention as to how the performance criteria of media are to be measured.

Driven by pressure from the pharmaceutical industry, media manufacturers have now become accustomed to providing certificates of analysis for each batch of media supplied. Clearly, these should be inspected on delivery and retained with the complete batch documentation.

### 2.8.2 *User's responsibility*

While accepting that the primary responsibility for quality rests with the manufacturer rather than the user of culture media, it is nevertheless recognized that, where the media are further processed, the end-user must accept responsibility for controlling any variation in that processing. The extent of testing will depend upon the extent of processing, which may range from heating to the addition of supplements.

The overall performance of a medium in use provides the final and complete quality assessment. However, the pharmaceutical microbiologist must provide independent evidence of the medium's quality characteristics and suitability for use before the product is released for use. Quality control tests carried out on the end-product in the laboratory should therefore confirm not only the medium's sterility and pH, but also its performance in productivity and selectivity tests, as well as its ability to produce distinctive colonies. As discussed previously, the importance of these tests has been recognized for some time in food microbiology through the publication of an international pharmacopoeia, detailing test methods, test strains and monographs on the most commonly used media in food microbiology (Baird *et al.*, 1985, 1987; Corry *et al.*, 1995). In pharmaceutical microbiology, however, the subject of quality control of culture media remains surprisingly a somewhat neglected area. With the exception of fertility tests carried out on sterility test media, the importance of performance tests has been largely overlooked until recently; in part this may have been due to the absence of pharmacopoeial requirements, but also in part to an unappreciated recognition of the significance of such tests and possibly excessive reliance on the commercial manufacturer. Recent editions of the *European Pharmacopoeia* (EP) (1997) and BP (1999) describe controls to demonstrate the nutritive and selective properties of media in the tests for specified micro-organisms; these tests do not, however provide the same degree of assurance as the performance and selectivity tests used for quality control purposes in food microbiology (see above). Many laboratories continue to depend upon fortuitous checks carried out during routine testing, rather than systematically testing the ability of the media to support or inhibit known test organisms.

The benefits of adopting such a standardized testing scheme, as outlined above, is that by using the proposed criteria and techniques to test media quality, the scheme may be used to:

- assess and compare the quality of commercially available dehydrated media or ready-to-use plates or tubes;
- check the quality of purchased batches of commercially available media, before use;
- audit the performance of media suppliers;
- check on medium preparation procedures.

Furthermore, the testing methods can be applied to the development and assessment of new media.

While recognizing the benefits of such a quality control scheme, the inevitable labour and equipment costs of implementing it are also acknowledged. Described below is an ecometric streaking method (Mossel *et al.*, 1983) which provides an easy and economic way of monitoring solid media quality on a routine basis, without incurring unjustifiable material and staff costs and increasing workloads.

### 2.8.3  *Ecometric streaking technique*

Overnight stationary broth cultures of test strains are inoculated onto the test and reference agar in a standardized way as shown in Figure 2.1. One loopful of inoculum only is sequentially diluted from streak to streak, with the loop being flamed and cooled in between each sector. Plates are incubated and growth on the plates is not recorded as a colony count, but as a score. Five streaks of growth in each quadrant score as one, while growth on up to three streaks score 0.5. A maximum score of 5 is obtained when all streaks in the four quadrants show growth and the final streak in the centre of the plate is also colonized. The sum of the score for each sector may then be calculated, giving the absolute growth index (AGI) for the test medium concerned; the relative growth index (RGI) for a given strain is the ratio of the AGI on

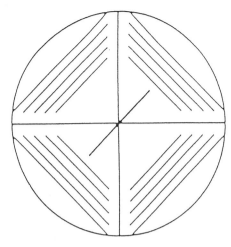

**Figure 2.1** The ecometric streaking template.

the test medium to the AGI on the reference medium. Thus, test and reference agars can be compared. It has been suggested that when ecometry is used for performance testing, RGIs not greater than 2.0 for unwanted strains and not less than 2.5 for wanted strains should be recorded (Corry *et al.*, 1995). It should be noted that the ecometric technique requires training in order to obtain reproducible, reliable results (Curtis, 1985; Weenk, 1995).

### 2.8.4 *Maintenance of test cultures*

A variety of test strains may be kept in the laboratory, including a minimum set of test strains used for quality control or testing purposes (e.g. preservative efficacy testing), a supplementary set used for more extensive testing, as well as wild strains which are individual to each laboratory. Organisms used for control purposes ideally should be strains from national collections. A system of stock culture maintenance such as that described in the Eighth Supplement to the *United States Pharmacopoiea* (1995) for preservative efficacy testing strains should be set up to preserve stock cultures, to minimize the risk of phenotypic changes, and to contain costs. This system should be documented and followed at all times.

### 2.8.5 *Control of ready-to-use agar plates*

Recent years have witnessed a considerable growth in the commercial supply of ready-to-use agar plates across the various microbiology disciplines, and specifically in pharmaceutical microbiology. There are several reasons for this, but among them the lack of skilled staff and adequate facilities, coupled with the requirement for cost containment, have driven microbiologists to seek alternative suppliers. Furthermore, the ability to cope with surges in demand has favoured the purchase of ready-to-use media. Pharmaceutical companies involved in sterile manufacture frequently require large numbers of settle plates to monitor environmental conditions; the convenience of using a pre-poured plate supply with freshly prepared media of known quality has obvious attractions for the user.

In terms of quality, although growth performance tests are of undoubted interest to the user, his or her prime concern is with the sterility of the batch which is to be introduced into critical production areas. Plate pouring operations inevitably involve aseptic processing, with its associated risk of failure. The use of irradiated plates may be justified for monitoring these critical operations.

The quality of poured plates deteriorates on storage, though with the use of appropriate packaging (both to protect from microbial contamination and to prevent the loss of moisture) and low-temperature storage, an acceptable shelf-life for the plate should be provided. It should be noted that the practice of incubating plates overnight at 35–37°C before use to confirm sterility reduces the growth performance characteristics of the medium and should be discouraged.

### 2.8.6  *External quality control schemes*

In food and clinical microbiology external quality control schemes provide an invaluable means of monitoring not only the performance of culture media but also the suitability of testing methods employed, as well as assessing the training and performance of staff involved in the testing. Uniform samples are sent from a central source to outlying laboratories and provide a way of measuring comparative abilities, often as part of a laboratory accreditation scheme. Ideally, these samples should be treated as routine samples and slotted blindly into the normal laboratory workload. Inevitably, as with broth filling runs, this is often not the case, leading to more care being taken with such samples. Pharmaceutical microbiologists have yet to set up such a centralized testing scheme, but the benefits are seen as considerable.

### 2.9  Trouble-shooting guide

From time to time, inevitably a batch of culture medium will fall short of its expected characteristics and some form of follow-up investigation will be required, initially within the laboratory and also possibly with the manufacturer at a later stage. If the problem persists, laboratory procedures should be checked thoroughly. Some of the likely problems and their possible causes, are listed in Table 2.2.

### 2.10  The future of culture media

During a century of use, culture media have been developed, adapted and refined to provide the nutrient requirements of a wide range of organisms. Different formulations, modifications to those formulations and fine adjustments have satisfied the microbial growth requirements of an increasing number of organisms of interest to today's microbiologist. In fundamental terms, however, culture media have changed very little in this period, the basic components having remained essentially the same. Culture media techniques provide a highly specific, sensitive, low-cost technology which is capable of detecting the replication of a single viable cell in both liquid and

**Table 2.2** Problems found in the manufacture of culture media. (Adapted from Oscroft and Corry, 1991.)

| Problem | Possible cause |
|---|---|
| Abnormal colour/darkening | Impure water |
| | Dirty glassware |
| | Deterioration of dehydrated medium |
| | Incorrect heating during sterilization |
| | Wrong pH |
| Coagulation | Medium too hot when supplements added |
| Flecks in media | Black: charred agar |
| | Clear: supplements added when agar too cool |
| Incorrect pH | Impure water or dirty glassware |
| | Overheating |
| | Chemical contamination |
| | pH meter calibration incorrect |
| | Faulty probe |
| | Reading taken at high pH |
| | Deterioration of dehydrated medium |
| Precipitation | Excessive heating |
| | Excessive storage ( >4 h) in molten state |
| | Deterioration of dehydrated medium |
| | Impure water or dirty glassware |
| | Medium too hot when supplement added |
| Reduced growth or effect on selective/differential properties | Incorrect weighing/mixing |
| | Contaminated water, glassware or tubing |
| | Deterioration of dehydrated medium |
| | Medium too hot when supplement added |
| | Medium too hot on contact with sample |
| | Prolonged medium storage |
| | Over-drying of plates |
| Soft agar | Excessive heating |
| | Acid hydrolysis of agar in low-pH media |
| | Inaccurate weighing/mixing |
| | Incompletely dissolved agar |
| | Incorrect volume of water |
| Toxicity | Excessive heating |
| | Deterioration of dehydrated medium |
| | Exposure to direct sunlight |
| | Wrong volume of supplement added |

solid media in a relatively short period of time. Through the use of broad-spectrum, non-selective media, unknown contaminants can be grown and identified.

Against this, however, the techniques are labour-intensive and require the input and interpretation of highly skilled, trained staff. Moreover, ever-increasing pressure is being applied to laboratory staff to provide test results in as short a time as possible. Large stocks of pharmaceutical products or of medical devices awaiting the results of sterility tests (a minimum 14-day period) can not only place a considerable strain on quarantine facilities but also represent an unacceptable delay in today's current economic climate where increasing demands are made on productivity and reduced costs. Alternative, new technologies have the attraction of offering rapid detection times without compromising the requirement for sensitive-specific methods. Such rapid methods are discussed in more detail in Chapter 7 and elsewhere (Denyer, 2000). It should be noted, however, that a move to rapid methods has itself placed great demands on microbiological media which are often employed in enrichment procedures.

## References

BAIRD, R.M., BARNES, E.M., CORRY, J.E.L., CURTIS, G.D.W. and MACKEY, B.M. (eds) (1985) Quality assurance and quality control of microbiological culture media. *International Journal of Food Microbiology*, **2**, 1–136.

BAIRD, R.M., CORRY, J.E.L. and CURTIS, G.D.W. (eds) (1987) Pharmacopoeia of culture media for food microbiology. *International Journal of Food Microbiology*, **5**, 187–300.

BRIDSON, E. (1994) The development, manufacture and control of microbiological culture media. Oxoid, Wade Road, Basingstoke RE24 8PN.

*British Pharmacopoeia* (1999) The Stationery Office, London.

BS 2646: Part 1 (1988) *Autoclaves for sterilisation in laboratories. Specification for design and construction.* British Standards Institution, London.

BS 2646: Part 2 (1990) *Autoclaves for sterilisation in laboratories. Guide to planning and installation.* British Standards Institution, London.

CORRY, J.E.L., CURTIS, G.D.W. and BAIRD, R.M. (eds) (1995) Pharmacopoeia of culture media. In: *Culture Media for Food Microbiology*. Elsevier, Amsterdam, pp. 243–491.

CURRY, A.S., GRAF, J.G. and MCEWEN, G.N. (1993) CTFA Microbiology Guidelines, The Cosmetic, Toiletry and Fragrance Association, Washington.

CURTIS, G.D.W. (1985) A review of methods for the quality control of culture media. *International Journal of Food Microbiology*, **2**, 13–20.

DENYER, S.P. (2001) Monitoring microbiological quality: application of rapid microbiological methods to pharmaceuticals. In: DENYER, S.P. and BAIRD, R.M. (eds), *The Guide to Microbiological Control in Pharmaceuticals and Medical Devices*. 2nd edn. Taylor and Francis, London, in press.

DENYER, S.P. and STEWART, G.S.A.B. (1998) Mechanisms of action of disinfectants. *International Biodeterioration and Biodegradation*, **41**, 261–8.

DEY, B.P. and ENGLEY, F.B. (1994) Neutralization of antimicrobial chemicals by recovery media. *Journal of Microbiological Methods*, **19**, 51–8.

DODD, C.E.R., SHARMAN, R.L., BLOOMFIELD, S.F., BOOTH, I.R. and STEWART, G.S.A.B. (1997) Inimical processes: bacterial self destruction and sub-lethal injury. *Trends in Food Science and Technology*, **8**, 238–41.

EISMAN, P.C., KULL, F.C. and MAYER, R.L. (1949) The bacteriological aspects of deionized water. *Journal of American Pharmaceutical Association Scientific Edition*, **38**, 88.

*European Pharmacopoeia* (1997) 3rd edn. EP Secretariat, Strasbourg.

GARLAND, A.J.M. (1999) A review of BSE and its inactivation. *European Journal of Parenteral Science*, **4**, 85–92.

HURST, A. (1977) Bacterial injury: a review. *Canadian Journal of Microbiology* **23**, 935–44.

MACFADDIN, J.F. (1985) *Media for isolation–cultivation–identification–maintenance of medical bacteria*. Vol. 1. Williams & Wilkins, Baltimore.

MOSSEL, D.A.A. and CORRY, J.E.L. (1977) Detection and enumeration of sublethally injured pathogenic and index bacteria in foods and water processed for safety. *Alimenta* (*Sonderausgabe*), 19–34.

MOSSEL, D.A.A., BONANTS-VAN LAARHOVEN, T.M.G., LIGHTENBERG-MERKUS, A.M.T. and WERDLER, M.E.B. (1983) Quality assurance of selective culture media for bacteria, moulds and yeasts: an attempt at standardisation on the international level. *Journal of Applied Bacteriology*, **54**, 313–27.

MOSSEL, D.A.A., CORRY, J.E.L., STRUIJK, C. and BAIRD, R.M. (eds) (1995) *Essentials of the Microbiology of Foods*. John Wiley, Chichester, pp. 63–83.

OSCROFT, C.A. and CORRY, E.J.L. (1991) Guidelines for the preparation, storage and handling of microbiological media. Campden Food and Drink Research Association. Technical manual no. 33.

PRESCOTT, L.M., HARLEY, J.P. and KLEIN, D.A. (eds) (1993) *Microbiology*, 2nd edition. William C. Brown, Dubuque, Iowa.

RAY, B. (1979) Methods to detect stressed micro-organisms. *Journal of Food Protection*, **42**, 346–55.

RUSSELL, A.D. (1991) Injured bacteria: occurrence and possible significance. *Letters in Applied Microbiology*, **12**, 1–2.

WEENK, G.H. (1995) Microbiological assessment of culture media: comparison and statistical evaluation of methods. In: BAIRD, R.M., CORRY, J.E.L. and CURTIS, G.D.W. (eds) *Culture Media for Food Microbiology*. Elsevier, Amsterdam, pp. 1–23.

*United States Pharmacopoeia* (1998) Eighth supplement to 23rd edn, Rockville: U.S. Pharmacopoeial Convention, pp. 4293–5.

**3**

# Sampling: Principles and Practice

ROSAMUND M. BAIRD

*Department of Pharmacy and Pharmacology, University of Bath, Claverton Down,*
*Bath BA2 7AY, United Kingdom*

## 3.1 General principles

Sample selection is the initial stage of a process whereby data on the characteristics of a batch are collected for evaluation. By definition, only a fraction of the batch is sampled for testing; clearly, therefore, that fraction must be representative of the batch in question. Since the fate of the batch depends upon the results generated from that first sample, sample selection must be regarded as a critical process and an essential part of the quality assurance system. As with any sampling process, appropriate predetermined indicators of quality, known as attributes, must be identified which reflect the characteristics of the batch; these are assumed not only to be homogeneously distributed throughout the batch but also to be recovered during a random sampling scheme. In microbiological sampling, however, these characteristics may not necessarily be randomly distributed throughout the batch and the sampling scheme should therefore reflect this. Sampling schemes currently employed in the pharmaceutical industry are increasingly based on an analysis of vulnerable points in a system, otherwise known as hazard analysis and control of critical points (HACCP), as discussed in more detail below.

Sampling for microbiological contamination poses additional problems. Owing to the risk of contamination, it should precede other forms of sampling. Previously unopened containers should be sampled, and suitably marked so that they can be easily identified in the case of failure. Additionally, special handling precautions should be adopted to minimize the risk of accidental contamination. These precautions will vary according to the sample type: when sampling an aseptically made product as part of a sterility test, specialized environmental conditions will be required (i.e. a Class A laminar flow cabinet in a Class B clean room, or an isolator located within a Class D clean room environment); however, for other less critical products, a dedicated sampling area in an uncontrolled environment may well suffice. Furthermore, since the microbiological characteristics of a given sample may change over time, account should be taken of this if sampling is carried out on a second occasion: localized environmental conditions may allow a contaminant to flourish in one situation but result in its death in another.

Sampling is an essential part of any quality assurance system. Standard operating procedures (SOPs) should therefore document how sampling is to be carried out (special precautions, methods and equipment), and who is authorized to take the samples. SOPs should be followed at all times. Personnel involved in sampling may not necessarily be microbiologists. They should, however, have been properly trained in the use of aseptic techniques and have received initial and regular on-going training to include: sampling plans; written sampling procedures; techniques and the use of dedicated equipment for sampling; the cleaning and correct storage of sampling equipment; the risks of cross-contamination; the precautions to be taken when sampling; the ease with which samples can be contaminated and hence invalidated; the importance of considering the visual appearance of materials, containers and labels; and the importance of recording any unexpected or unusual circumstances.

Additionally, SOPs should document the quantity of sample to be taken, how it is to be subdivided, the type of container to be used, and particular storage conditions. The interval between taking the sampling and testing should be minimal, particularly when sampling water supplies. If left at ambient temperature, contaminants may multiply during this time, leading to erroneously high counts. Gram-negative opportunist pathogens, such as *Pseudomonas* spp., have minimal nutritional requirements and may cause particular problems in stored water samples. If necessary, samples should be refrigerated, maintaining a temperature of less than 7°C, and preferably 1–4°C. If testing is not performed on site, or if there is to be a delay, it may be necessary to store samples in insulated boxes containing ice cubes, or to use specialized transport media. These will retain the viability of any contaminants but prevent their multiplication during this period.

## 3.2  Sampling guidance

In setting up any sampling scheme, it is first necessary to decide on the required level of assurance. Clearly the more samples that are taken and the larger the sample size, the greater the assurance of accuracy of the results. British Standard published tables (BS 6001, 1972) provide guidance on sampling rates and levels of assurance; a dramatic increase in sample numbers, however, is required to provide increasing assurance. A fine balance may need to be struck between what is regarded as an acceptable level of assurance, taking into consideration all quality indicators in a defined process, and the maximum number of samples which can be drawn from the batch to provide this level of confidence. Over a period of time sufficient confidence may be built up in batch quality that reduced sampling schemes may be justified. This may involve a reduction in the number of samples per batch or the number of batches tested, or even the number or types of tests carried out on each sample. Although decreasing the number of samples will affect the discrimination of the test, this may be of academic importance if the batch size is large in comparison with the sample size. Testing only one batch in five for example may be justified from accumulated results over a period of time. However, should test results indicate a deterioration in product quality, sampling plans can be switched from normal inspection to a tightened inspection (i.e. every delivery), as stated in BS 6000 (1972) and BS 6002 (1972), until test results show that normal inspection levels are again warranted.

### 3.2.1  *Single sampling*

Traditional sampling methods have been based upon the analysis of a single composite sample taken from selected containers in the delivery, invariably the square root of the number of containers plus two. A portion of this sample is analysed, and the delivery is then accepted or rejected on the basis of the result. The statistical basis of this sampling scheme is unknown (Russell, 1996), but if necessary the composite sample can be easily resampled and results of individual products can be examined for trends over a period of time.

### 3.2.2  *Attribute sampling*

An alternative method of sampling, widely adopted in the cosmetic and food industries, involves the selection of attributes which must be met in full or part by a predetermined number of samples taken from the delivery. By definition, such schemes incorporate so-called tolerances, so allowing a small proportion of samples to show slight deficiencies; however, clearly such schemes are unsuitable for the testing of sterile products. Samples are taken as before but analysed separately, giving an increased level of assurance over traditional methods. Although requiring more resources in terms of time and materials, attribute sampling can be regarded as more effective if properly targeted to vulnerable materials.

#### Two-class attribute scheme

In the two-class attribute sampling plan the number of samples and the maximum permitted number of positive results are defined (*n* and *c* respectively). Such a scheme is used in the bacteriological testing of drinking water supplies (Anonymous, 1982). From five samples taken (*n*), *Escherichia coli* is not permitted in any sample and only two of the five samples may contain Enterobacteriaceae. Thus, two quality levels are defined by the required absence of *E. coli* and by the limited number of samples which may contain Enterobacteriaceae. In pharmaceutical terms this scheme is suitable where positive results are unacceptable. For example, a raw material of natural origin, such as thyroid powder, may have a specification requiring the absence of salmonellae in 25 g (where *n* is $5 \times 5$ g and *c* is 0).

#### Three-class attribute scheme

In this scheme three quality levels are defined: the fully acceptable, often known as the acceptable quality level or AQL; the marginally acceptable; and the unacceptable. Widely used in the food and cosmetic industries, these schemes have been closely associated with the establishment and use of microbiological reference values (Mossel *et al.*, 1995; CTPA 1996). Their use in pharmaceutical microbiology has been mainly restricted to the in-use testing of raw materials and non-sterile products, where viable counts form the basis of acceptance or rejection; they have yet to be incorporated into pharmacopoeial specifications. In introducing an intermediate level, trends in results thus become more visible; only the intermediate values need to be recorded, since the other values are either acceptable or unacceptable. For example, a viable count of less than 100 colony forming units (c.f.u.) may be regarded as acceptable, while a count of more than 1000 c.f.u is unacceptable; counts between these two values may be regarded as marginally acceptable in a defined number of samples. Such a scheme was adopted by the Cosmetic, Toiletry and Perfumery Association (CTPA) in their 1990 Guidelines of microbial limits for products in their final container at the time of manufacture. From five randomly selected units, no more than two samples were allowed to fall into this intermediate level. Should this number be exceeded, the course of action to be taken was set out: production was to be suspended until the cause was ascertained; the batch was to be held and tests repeated on the five initial samples; and remedial action was to be taken. In addition, 20 more units should be sampled and tested for a total viable count (TVC) and the presence of harmful organisms. If all results were then found to be below the upper limit and no harmful organisms were detected, the batch could then be released, but otherwise the batch should be rejected. In addition, subsequent batches should be sampled at a level of 20 samples per batch until results showed that the remedial action had been effective.

### 3.2.3 *Resampling*

The need to resample requires careful consideration. If there is any doubt as to the validity of sampling, a second sample must always be taken. When the accuracy of the counting method has been questioned, or perhaps a non-homogeneous distribution of micro-organisms is suspected, a re-examination may be justified. Microbial populations may be dynamic, changing over a period of time and reveal a different microbial profile on a second sampling; initial contaminants may well have died, thereby allowing secondary contaminants to flourish. Where unacceptable organisms are found, there is no place for a re-test. Clearly defined requirements for retesting sterility test samples are now found in the pharmacopoeias (see section 3.3.1).

### 3.2.4 *Reference samples*

Clearly, samples taken as reference material should be representative of the batch from which they were taken. They should be suitably labelled, indicating the reference number, the contents, the batch number, the date of sampling and the containers from which the samples were withdrawn. Reference samples should be retained from each batch of finished product until one year after the expiry date. They should also be of a sufficient size to permit at least a full re-examination. Appropriate precautions should be taken to prevent damage to samples during storage which would invalidate the results.

### 3.2.5 *HACCP approach to management of quality*

The maxim that patient protection can only be achieved by effective measures of intervention, and not by end-product testing alone, is not new. The notion that quality should be built in at every stage of manufacture was implicit in the first *Guide to Good Manufacturing Practice* (1971), but was also widely appreciated and accepted by the pharmaceutical industry at this time. Since then, ecological in-house studies have provided a wealth of information, albeit mainly unpublished, not only on the microbiological flora normally associated with various raw materials but also on the impact of processing, product formulation, storage and distribution conditions on that flora. Such information, combined with the design specifications of manufacturing environment and processing equipment has enabled the microbiologist to predict the probable hazards associated with the product and to identify the likely points for microbial contamination or growth. Once these critical control points have been identified and estimated at each step in the manufacturing process, they should be monitored regularly and intensively; if necessary, appropriate corrective steps should also be taken to ensure that control is maintained at all times. Controls include three lines of defence: (i) the limitation of contamination by selecting suitable raw materials and applying appropriate hygienic measures during manufacture; (ii) minimizing all opportunities for microbial growth throughout product manufacture, distribution and storage; and (iii) processing for safety (i.e. sterilization) if a safe product cannot be ensured by the two previous lines of defence. The concept of hazard analysis and control of critical points (HACCP) does not end with the manufacture of a quality finished product, but it should be extended longitudinally throughout the storage and distribution chain until its eventual use by the patient. This total quality concept has been known as Longitudinal Integrated Safety Assurance (LISA) in the food industry for some time (Mossel *et al.*, 1995).

Nowadays, in both the pharmaceutical and medical device industries the HACCP concept is an essential part of any quality system. Control strategy is therefore managed prospectively by anticipating and addressing the potential risks, rather than reacting to them when they have become a reality. Hazard analysis forms an integral part of any validation exercise in identifying where a process can go wrong, and by the same token how the process can be controlled. Through its adoption, an improvement in the microbiological profile of products should be seen with fewer fluctuations in quality; in addition any trouble spots should be easily identified and quickly resolved.

From the above discussion, the effect of HACCP on sampling theory is therefore seen in the adoption of a system in which vulnerable points in a process are targeted, rather than a random selection of samples. By analysing these critical samples, invaluable information is thus generated not only on the extent of control within the system but also whether there are any significant fluctuations in quality. In addition, HACCP can also be used to validate the effectiveness of proposed hygienic procedures. Using a bracketing technique of analysing samples before and after a particular process, the effectiveness of that procedure can be clearly demonstrated, i.e. whether the microbial load is affected in any way.

## 3.3  Specialized sampling

Sampling schemes should clearly reflect the nature of the product.

### 3.3.1  *Sterile products*

When sampling for sterility, the level of assurance concerning the quality of the batch is a function not only of the homogeneity of the manufacturing conditions but also of the efficiency of the sampling plan. The time-honoured test for sterility, as described in the pharmacopoeias, provides microbiological information on the sample tested under specified conditions; sample selection is critical, and herein lies one of the major deficiencies of the test. By its nature the test is destructive and by definition it is applied only to a random selection of statistically drawn units. The statistics of sampling for sterility have been discussed at length by Brown and Gilbert (1977).

In mathematical terms, the proportion of sterile and non-sterile containers in a batch may be defined as $q$ and $p$ respectively, such that:

$$p + q = 1 \quad \text{or} \quad q = (1 - p)$$

Taking $n$ samples from the batch for testing, the probability $P$ of all of these samples being sterile and the batch therefore passing the test is:

$$P = q^n \quad \text{or} \quad P = (1 - p)^n.$$

Supposing 1% of containers in the batch were contaminated, i.e. $p = 0.01$, and ten samples were taken for testing, then the probability $P$ of passing the test is:

$$P = (1 - 0.01)^n \quad \text{or} \quad P = 0.904$$

In practice, this means that in taking ten samples from this non-sterile batch, the batch will in fact pass the test on nine out of ten occasions when tested. Depending on the incidence of contamination and the number of samples taken, the probability ($P$) of passing the sterility test will vary, as shown in Table 3.1.

It is assumed that the entire contents of each sample container have been used for the test, but this may not necessarily be the case. The 1999 *British Pharmacopoeia* (BP) for example states that for parenteral products of less than 1 ml, the entire contents of the container should be sampled, whereas for those containing 1 ml or more, only 50% of the contents of the container need to be sampled. On the other hand, *United States Pharmacopoeia* (USP) XXIII (1995) requires the entire contents of parenteral products of less than 1 ml and of 50 ml and over to be sampled for testing; for intermediate volumes, a proportionate sample is permitted (1 ml for volumes less than 10 ml, 5 ml for volumes between 10 ml and less than 50 ml). Clearly, such testing schemes can have a profound effect on the results of samples with low levels of contamination; contamination may simply not be detected. The implications of this are well illustrated in Table 3.2.

**Table 3.1** The probability of a batch passing a sterility test for various degrees of contamination and sample sizes

| Sample size | Percentage of contaminated items in the batch | | | | |
|---|---|---|---|---|---|
| | 0.1 | 0.5 | 1.0 | 3.0 | 5.0 |
| 10 | 0.99 | 0.96 | 0.91 | 0.74 | 0.60 |
| 20 | 0.98 | 0.90 | 0.82 | 0.54 | 0.35 |
| 50 | 0.95 | 0.78 | 0.61 | 0.22 | 0.08 |
| 100 | 0.91 | 0.61 | 0.37 | 0.05 | 0.01 |

**Table 3.2** The probability of a batch passing a sterility test against the sample volume taken from each test container

| Sample volume (ml) | Probability of batch passing the test ($P$) |
|---|---|
| 10 | 0.998 |
| 20 | 0.996 |
| 100 | 0.980 |
| 200 | 0.960 |
| 500 | 0.904 |
| 1000 | 0.817 |

The container sample size is 20 ($n = 20$), the volume in each container is 100 ml, the incidence of contamination is 1% and the level of contamination is 1 c.f.u. $l^{-1}$

Supposing that 1% of 1-litre containers in a batch are contaminated, i.e. $p = 0.01$, and that those defective units contain 1 c.f.u./1000 ml, the calculated probability ($P$) of passing the sterility test, based on 20 sampled units ($n$), will vary according to the sample volume examined. As the sample volume increases, so there is less chance of accepting the defective batch. Use of a membrane filtration technique for testing offers obvious advantages over a direct inoculation method, since in theory at least the sample volume under test is not restricted to small quantities.

In mathematical terms, it follows that the probability of rejection can be expressed as $1 - (1 - p)^n$. In the event of a sterility test failure, and if the test is subsequently shown to have been invalid according to defined criteria, the batch may be resampled once taking the same number of units as in the original selection [*European Pharmacopoeia* (EP) 1997; BP, 1999]. From a statistical point of view, the probability of approving a defective batch in fact increases when the batch is sampled on this second occasion, since a proportion of the batch has already been removed. In mathematical terms, it can be expressed as:

$$1 - p^n[2 - (1 - p)^n]$$

Clearly, the chance of detecting contamination increases with the number of contaminated units in a batch. Guidance on the number of units to be sampled in relation to the batch size is given in current pharmacopoeias (USP, 1995; EP, 1997; BP, 1999), as are the quantities of product to be sampled from each container.

For heat-sterilized products, samples should be taken from specified parts in the load, previously identified during validation studies as the slowest to reach sterilizing temperatures. Here clearly the risk of sterilization failure is deemed to be highest in those samples taken from the coolest part of the load. For aseptically made products, both random and non-random sampling is required; one-quarter of the required samples should be taken from the beginning of the filling run and similarly from the end of the run, the remaining half being selected randomly from the finished batch. Where significant interventions have interrupted the filling process, samples should additionally be taken immediately after this event. Samples for pyrogen testing should be collected from among the last-filled containers on the basis that any endotoxin production is likely to be greatest in the unfiltered bulk which has been left standing at ambient temperature during the filling period. Sampling of sterile products is also discussed in Chapter 8.

## Process simulation in aseptic manufacture

The microbiological status of an aseptic manufacturing process must be qualified during validation work through process simulation, or broth filling trials, whereby the filling process itself is simulated using nutrient broth in place of product. Broth filling should follow the routine aseptic manufacturing process as closely as possible and include all the critical manufacturing steps, planned interventions and the normal complement of staff. Process simulation should be repeated at defined intervals (often 6-monthly) and after any significant modification to equipment or process. Any lapse in aseptic manufacturing is shown by the appearance of visible growth (turbidity) in the broth following incubation.

At least three successful consecutive filling trials are required to validate a new aseptic process, incorporating all product shifts for each line/product/container combination. The validation programme should consider issues such as the method of fill, fill volume, choice of media, incubation conditions, inspection procedures, documentation, interpretation of results and corrective action. The number of sample units to be filled is a critical issue: traditionally 3000 units has been accepted as the minimum number to detect a contamination rate of 0.1% at a 95% confidence level and thus provide a valid evaluation. Many pharmaceutical companies nowadays fill well in excess of this number at one time, and can show contamination rates well below 0.1%. For clinical trial manufacture using pilot plant facilities, smaller broth fills are acceptable, but they should at least equal the size of the product batch. More detailed information is contained in publications by the United States Food and Drug Administration (1987) and the Parenteral Society (1993).

### 3.3.2  *Bioburden sampling*

Aqueous products are commonly assumed to be homogeneously contaminated; nevertheless, the contents of individual containers should be mixed thoroughly before samples are withdrawn. In recent years increasing attention has been paid to the occurrence of biofilms and their significance, even within a simple aqueous product. Where surface biofilms have formed, contamination levels may be inconsistent within a product; the biofilm may slough off and discharge erratically, then reform, and during this time contamination levels may rise and fall dramatically. The possible existence of biofilms clearly presents sampling problems. In practice, however, they are most likely to be encountered in piped water supplies, particularly if the flow is stagnant. Piped water is often circulated at $1-2\,\mathrm{m\,s^{-1}}$ to prevent biofilm formation.

In non-aqueous products, pockets of contamination may occur; detection of this contamination may simply be a matter of chance. Aerobic contaminants may grow preferentially on the surface of topical products. Sampling schemes should clearly take account of this difference in distribution. In one notable example surface contamination of an antibiotic eye ointment was reported to have occurred through the condensation of water onto the ointment surface during manufacture, thereby enabling local proliferation of *Pseudomonas aeruginosa* in an otherwise hostile environment (Kallings *et al.*, 1966). Use of this ointment resulted in loss of sight for eight individuals. Quality control tests had failed to detect the contamination owing to inadequate sampling methods. Oily products, creams and ointments frequently present sampling problems. Contaminants trapped within the matrix of the product will fail to grow until released from their immediate environment. The BP (1998) and USP (1995) recommend the use of a suitable emulsifying agent, such as 0.1% (w/v) polysorbate 80. The resulting homogenate can then be warmed to not more than 40°C for a maximum of 30 min. While this defined period may restrict the multiplication of contaminants, it should not be forgotten that preservative action is temperature-dependent and this elevated temperature could have a deleterious effect on any indigenous microbial flora, thereby providing misleading information on product quality. Sampling efficiency may also be improved in oily products by combining increased temperature with the use of a blender or homogenizer, such as a Stomacher (Colworth). When sampling aerosols, these should be sprayed directly into a tared vessel containing diluent; if necessary, a suitable quantity of antifoam may be added to break the foam emulsion.

Sampling programmes set up to determine the bioburden in raw materials, intermediate or finished goods will inevitably concentrate on those products which are known to be susceptible to microbial contamination or have a demonstrated microbiological history. Sampling schemes for individual products will be affected not only by the nature of the manufacturing process and the batch size but also by the antimicrobial activity of the formulation. Guidelines adopted by the UK cosmetic industry advocate that samples should be taken from the beginning, middle and end of each day's production (CTPA, 1996). Sampling plans should be reviewed at least annually.

### 3.3.3  Water sampling

Sampling of water supplies is governed by its usage, and its method of preparation and storage. If mains water is used for any production purposes, it should be sampled at least weekly. Deionized water should be sampled weekly or between regeneration cycles of the deionizing bed, whichever is the shortest. During commissioning and validation of the system, it should be established that there is no build-up of unacceptable bacterial contamination before regeneration occurs. Distilled water for both parenteral and non-parenteral manufacture should be sampled frequently to ensure that microbiological control is tightly maintained.

In drawing up a sampling plan for any water treatment and distribution system, a detailed schematic of the system should be consulted and kept available for routine audits and inspections. Samples should be taken from every point of treatment (e.g. post deionization, post reverse-osmosis, still) and every outlet to ensure that the quality of water at every processing, distribution and storage stage is monitored; in the event of unacceptable contamination levels, the source of contamination will be more easily identified.

Water systems present more problems than others in terms of what constitutes a reasonable sample size. Clearly, the higher the expected water quality, the lower the expected TVC, and the greater the sample volume must be to confirm this with any degree of confidence. In practice, an acceptable sample size usually represents a mere fraction of the total capacity of the system; however, on occasions the subject of sample volume is not given due consideration, and ridiculously small sample volumes may be encountered in the course of auditing exercises. FDA guidance (Anonymous, 1976) proposed that the sample size should vary with the water type: for water used in cleaning and rinsing, three consecutive samples of 100 ml should be taken; for water used in manufacture or final rinsing of equipment and containers, three consecutive samples of 250 ml or more should be taken from the same site; for water used in cooling the product after sterilization, three consecutive samples of 1 litre or more should be taken from the same site. Water for endotoxin testing requires a sample volume of less than 1 ml.

Sampling may be via take-off points with taps or via specialized in-line sanitary sampling devices. Meticulous attention should be paid to the selection, siting, maintenance and cleaning of these outlets. Stagnant water left in these outlets can only too easily become a focus of contamination; this can affect not only the quality of the water sample itself but, more importantly through a process of back-tracking, may also affect the quality of the water supplies. Once established as a source of contamination, it may prove to be very difficult to eradicate. The SOP should state that water must flow freely from each outlet for 2 min before the sample is collected into a sterile container. Some microbiologists may also insist upon flaming the outlet before sampling. Samples for endotoxin assay should be collected in depyrogenized containers.

### 3.3.4 *Environmental sampling*

Sampling methods used to monitor environmental contamination levels should not interfere with zone protection. In sampling for airborne levels of contamination, the use of settle plates offers clear advantages over more cumbersome active sampling techniques, such as the slit sampler. The design of these instruments has, however, improved in recent years and in some instances it is possible to use a remote sampling probe in critical areas, without the necessity of having to bring in bulky items of equipment. Additionally, active sampling techniques provide a quantitative assessment of the level of contamination/unit volume of air, either utilizing an impaction method of capture (as in the case of the Casella or New Brunswick slit sampler, the Cherwell slit to agar surface air system (SAS) sampler, sieve samplers or the Millipore MD8 gelatin pad sampler), or alternatively a centrifugal sampling method (RCS Biotest), or less commonly a liquid impingement method. The advantages and limitations of these techniques are summarized in Table 3.3.

While all these techniques sample relatively large volumes of air for a set time, they provide only a 'snap-shot' picture of clean room activities at a given point of time, and give no indication of conditions before or after sampling. Each sampling method has its limitations, and these should be understood when sample data are being evaluated, since the results may not be directly comparable (Ljungqvist and Reinmüller, 1998). The results should therefore be interpreted as providing an indication of airborne contamination, rather than as a true absolute value.

In contrast, passive sampling techniques simply provide an indication (i.e. semi-quantitative assessment) of the microbial fall-out from the atmosphere, albeit over an extended period of time and activity. The technique is affected by local air currents, exposure times and the particle settling rate. According to Stokes' law, a 10 μm particle has a settling rate in still air of 0.6 cm s$^{-1}$. Sensitivity of the technique can be improved by increasing the exposed surface area and the exposure time. Plates may safely be exposed for at least 4 h without encountering problems of dehydration. Besides their

**Table 3.3**   Comparison of various air sampling methods

| Method | Sample volume $(1\ min^{-1})$ | Advantages | Limitations |
|---|---|---|---|
| Slit sampler | 50–700 | Large air sample Time-related contamination Good particle retention | Air disturbance Cumbersome Intrusive |
| Surface air system | 180 | High capture efficiency | Low sampling rate generates particles High air velocity can reduce cell viability |
| Gelatin pad sampler (Millipore MD8) | 133 | High capture efficiency | Fragile gelatin filter Colony counting problems |
| Biotest | 40 | Portable Convenient | Efficiency depends on particle size |
| Air liquid impinger | 28.3 | High capture efficiency | Additional handling, sampling and counting |

convenience, settle plates provide a simple and cheap way of monitoring aerial contamination levels, particularly in large units. In essence therefore these two sampling techniques provide different but complementary information.

Sampling schemes for environmental monitoring of controlled areas are employed during commissioning of the unit in the unmanned state, and also during periods of operator activity. Then there is the opportunity to identify areas of air flow turbulence, stagnation, and operator activity where future sampling should be concentrated. Critical processing zones in aseptic areas should be sampled during each operational production shift, particularly where the product is exposed to the environment at the point of fill; other processing zones should be monitored daily when aseptic processing occurs. As discussed previously, for other areas sampling schemes should not be cast in tablets of stone but should be reviewed on a regular basis. Provided that the required level of control is maintained, it is indeed possible that the frequency or intensity of sampling may be reduced. However, should results indicate that conditions have deteriorated, then the sampling plan should revert to what amounts to a tightened inspection regimen until a reduced sampling scheme can again be justified.

The effectiveness of cleaning and disinfection policies on hard surfaces (walls, floors, worktops) can be assessed using surface sampling methods, such as Rodac or contact plates or swabs. Account should be taken of disinfectants in current use; appropriate inactivation methods should be used so that antimicrobial activity is not carried over to the media (see Chapter 2). Microbial adhesion to surfaces is affected by a number of factors: the surface type; contaminant involved; the presence of organic matter; and the source and method of contamination. These in turn will affect the efficacy with which micro-organisms can be recovered using the chosen sampling method. Contact plates (55 mm diameter) generally have good transfer efficiency, provided there is microbial adhesion to the agar and that the single cell isolated is capable of producing a colony on the given medium. Their use, however, is restricted to flat smooth surfaces, and afterwards residual medium must be removed. Swabs on the other hand can be used to sample relatively large irregular areas or inaccessible sites; variables include not only the actual surface wiping method, but also the transfer efficiency between surface and swab and then between swab and agar. Chemicals impregnated into the swab itself may also affect survival and growth of isolates. Environmental sampling methods are discussed further in Chapter 8.

### 3.3.5 *Cleaning materials*

The sampling of cleaning equipment for contamination has traditionally been a much neglected area; the notion that such equipment can not only be a potent source of contamination but also be responsible for the spread of contamination is not widely appreciated among those responsible for cleaning (see Chapter 12). However, simple microbiological demonstrations of contaminated samples and appropriate training courses can quickly dispel this lack of understanding, leading to a greater sense of commitment concerning contamination control issues and the part that individuals can play in promoting improved environmental control. Samples of cleaning materials may contain remnants of disinfectants or cleaning agents which will require inactivation before testing, as discussed in Chapter 2. The effectiveness of the proposed method should be validated.

### 3.3.6 *Equipment*

Microbiological sampling of equipment is carried out not only to validate the efficiency of proposed cleaning methods but also to monitor those methods on a routine basis. Sampling methods may need to be ingeniously adapted in order to have a reasonable chance of recovering contaminants located on equipment. Flushing equipment with sterile rinse solutions which are then filtered is a commonly used method. Some form of agitation or stirring may increase recovery rates. Large surface areas can be examined, as well as inaccessible parts of equipment that cannot be routinely disassembled for evaluation. Levels of contaminants as well as product residues can then be measured directly in rinse solutions. Recent guidance has been published in draft form on defined criteria which product residues must meet in order to find acceptance (Anonymous, 2000). Alternatively, direct surface sampling with swabs may be used to transfer contaminants from the equipment surface to the agar plate. The ability to recover contaminants may be affected by the choice of sampling material as well as the sampling medium. This semi-quantitative method is generally regarded as less efficient at recovering surface contaminants than the above method; a combination of both methods probably provides the best solution. It cannot be assumed that contamination will be evenly distributed throughout; pockets of contamination may well occur, particularly if design faults exist, e.g. dead legs, inside threads or areas where biofilms may become established. By the same token, it cannot be assumed that contamination would be evenly recovered.

### 3.3.7  *Operator sampling*

Contact plates may be used to sample directly contamination levels on clean room clothing, including gloves. Finger dabbing exercises where hands are lightly pressed on the agar surface for a few seconds will not only provide information on the effectiveness of hand-washing and gloving techniques but also give an indication of staff discipline in the unit. An alternative method of sampling contamination on gloved hands involves washing hands in a sterile buffer or rinse solution, followed by filtration and incubation of any contaminants retained on the filter.

### 3.3.8  *Medical devices*

In contrast to raw materials used in pharmaceutical production, the component materials of medical devices are not normally tested for microbial contamination on a routine basis. However, bioburden estimates may be necessary on suspect raw materials where a high bioburden has been found prior to sterilization of the manufactured device. Likewise, with sterile devices little significance is attached to sterility testing of the final product; rather the concept of sterility assurance has been based upon proper control of manufacturing and sterilization processes and the satisfactory completion of all associated microbiological controls. On occasions, however, where devices have been aseptically processed, microbiological testing of the final product may be required; alternatively, it may form part of a validation exercise where there is to be a reduced exposure to the sterilizing agent.

In contrast, considerable significance is attached to bioburden testing which is regarded as a critical step in the manfacturing process of sterile devices. Reliable, accurate and reproducible data must therefore be collected. Any under-estimation of the bioburden population could result in a miscalculation of the sterilization requirements for a given product; whereas an over-estimation could result in excessive exposure to the sterilizing agent, which in turn could affect the functioning of the device. In essence, bioburden estimates form part of the overall quality system in manufacturing medical devices, providing supportive data on: the microbiological quality of incoming materials and how storage conditions may have affected these components; the effect of processing on microbial quality of the product; the efficacy of the cleaning process; and the usefulness of the environmental programme.

Detailed guidance on bioburden estimation in medical devices is given in EN 1174 Part 2 (1994). Five distinct stages are involved: sample selection; removal of micro-organisms from sample; transfer of micro-organisms to recovery conditions; enumeration of micro-organisms with specific characteristics; and interpretation of data, involving application of appropriate correction factors determined during validation studies. Medical devices comprise a diverse collection of product types and there is no single sampling technique which can be applied to this range of products. Figure 3.1 shows the considerable variation in bioburden which may be found on different types of medical devices, and well illustrates the sampling problems. Furthermore, it is not possible to determine the exact level of microbial contamination; the 'immediate' bioburden following manufacture will differ in magnitude and variety from the 'retained' bioburden found on the product after a period of storage.

In sampling for pre-sterilization counts, items should ideally be selected at random from the production line just prior to sterilization. Otherwise, a finished but unsaleable

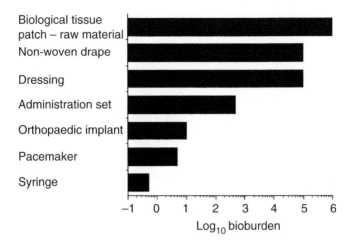

**Figure 3.1** Magnitude of bioburden on devices. (Based on Hoxey, 1993.)

product may be selected, provided that it has been exposed to the complete manufacturing process and it represents the bioburden of the batch. The entire product should be sampled whenever possible; otherwise a representative portion should be selected for sampling, assuming an even bioburden distribution. If this is not the case, the location of the highest microbial population should be sampled. Large items may need to be cultured directly, using contact plates, swabs, an overlay technique using molten agar at 44°C, or using a direct transfer to liquid media. Sampling should take place in conditions where the risk of accidental contamination is minimized, i.e. in a laminar flow air flow cabinet using suitably clad staff who have been trained in aseptic manipulations.

Sample size and frequency will clearly be greatest when producing a new product. Numbers may be gradually reduced, but should nevertheless remain sufficiently frequent to allow changes in bioburden due to seasonal variation, manufacturing change or material change to be detected. There should be a minimal delay between sampling and testing. Where delays are inevitable, the maximum storage time and conditions should be stated.

Sampling problems may well be encountered in the course of bioburden estimation. Microbial cells may be embedded in, or attached to, the surface of the device and may require some form of physical treatment to release them. The degree of microbial adhesion to a given surface varies according to the individual microbe, surface type and the presence of any other materials, for example lubricants. Removal treatments may involve rinsing, combined with some form of physical force or direct surface sampling. Where surfactants are included to enhance recovery, they should not be inhibitory to microbial growth, particularly at higher concentrations and should not create excessive foaming with consequent loss of contaminants to the foam layer. In selecting a suitable sampling method, the following requirements should be met:

- it should not introduce contaminants;
- it should not stress the microbial population (for example through shear forces, temperature rise or osmotic shock);
- it should provide reproducible results.

**Table 3.4**  Sample treatment methods in bioburden estimation

| Method | Defined parameter | Advantages/use | Limitations |
| --- | --- | --- | --- |
| Stomaching | Time | Absorbent items | Items causing bag puncture |
| Ultrasonication | Time<br>Loading in bath<br>Sonication frequency | Impermeable items with complex configuration | Limited throughput<br>Cell disruption with excessive treatment |
| Shaking with glass beads | Time<br>Frequency | Straightforward technique | Not necessarily effective for lumens |
| Vortex mixing | Container type<br>Time and speed<br>Manual pressure | Quick and simple | Uncomplicated small samples with regular surfaces |
| Flushing | Contact time<br>Rate of flushing<br>Eluent volume | Tubing – particularly if only fluid path is to be examined | |
| Disintegration | Time | Non-woven fabrics | Eluent overheating<br>Possible cell damage |

Recovery systems can markedly affect the overall efficiency of the method used. Various diluents and eluents may be used, e.g. sodium chloride (0.25–0.9%), one-quarter strength Ringer's, peptone water (0.1–1%), buffered peptone water, phosphate-buffered saline, or Calgon Ringer's solution. Careful selection is required so that diluents' inherent properties (chemical nature, concentration, toxicity and pH) do not affect microbial recovery. Care should also be taken in the choice of added surfactants (e.g. 0.01–0.1% Tween 80 used for extracting micro-organisms from solid surfaces). Where there is a possibility that toxic substances may have been released from the device into the eluent, appropriate treatments must then be devised to neutralize or separate out the cells by physical means. The efficiency of this procedure must also be validated. Sample treatment methods used in bioburden estimation are summarized in Table 3.4.

Finally, the possibility that injured cells may exist within any surviving microbial population cannot be discounted. These cells may require suitable resuscitation conditions for the repair of cell damage and before further growth will take place, as discussed in more detail in Chapter 2.

## References

Anonymous (1976) Proposed rules. Current good manufacturing practice in the manufacture, processing, packaging or holding of large volume parenterals. *Federal Register*, **41**, No. 106.
Anonymous (1982) *The Bacteriological Examination of Drinking Water Supplies*. HMSO, London.
Anonymous (2000) Validation master plan design qualification installation and operational qualification, non-sterile process validation cleaning validation. Draft Annex 15 to the EU Guide to Good Manufacturing Practice. Volume IV, In press.
*British Pharmacopoeia* (1999) The Stationery Office, London.

British Standard 6000 (1972) *The use of BS 6001, sampling procedures and tables for inspection by attributes*. British Standards Institution, London.

British Standard 6001 (1972) *Sampling procedures and tables for inspection by attributes*. British Standards Institution, London.

British Standard 6002 (1972) *Sampling procedures and charts for inspection by variables for per cent defective*. British Standards Institution, London.

BROWN, M.R.W. and GILBERT, P. (1977) Increasing the probability of sterility of medicinal products. *Journal of Pharmacy and Pharmacology*, **29**, 517–23.

CTPA (1990) Microbial quality management – CTPA limits and guidelines. The Cosmetic, Toiletry and Perfumery Association, Ltd, London.

CTPA (1996) Microbial quality management. The Cosmetic, Toiletry and Perfumery Association, Ltd, London.

BS EN 1174 (1994) Sterilization of medical devices – Estimation of the population of micro-organisms on product. British Standards Institution, London.

*European Pharmacopoeia* (1997) Maisonneuve. Third edition.

*Guide to Good Manufacturing Practice* (1971) HMSO, London.

HOXEY, E. (1993) Validation of methods for bioburden estimation. In: MORRISEY, R.F. (ed.), *Sterilization of Medical Products*. PolyScience, Movin Heights, Canada.

KALLINGS, L.O., RINGERTZ, O., SILVERSTOLPE, O. and ERNERFELDT, F. (1966). Microbiological contamination of medicinal products. *Acta Pharmaceutica Suecica* **3**, 219–30.

LJUNGQVIST, B. and REINMÜLLER, B. (1998) Active sampling of airborne viable particles in controlled environments: a comparative study of common instruments. *European Journal of Parenteral Science*, **3**, 59–62.

MOSSEL, D.A.A., CORRY, J.E.L., STRUIJK, C.B. and BAIRD, R.M. (Eds) (1995) In: *Essentials of the Microbiology of Foods*, pp. 219–223. John Wiley & Sons, Chichester.

Parenteral Society (1993) The use of process simulation tests in the evaluation of processes for the manufacture of sterile products. Technical Monograph No. 4.

RUSSELL, M. (1996) Microbiological control of raw materials. In: BAIRD, R.M. and BLOOMFIELD, S.F. (eds), *Microbial Quality Assurance in Cosmetics, Toiletries and Non-Sterile Pharmaceuticals*. Taylor & Francis, London, pp. 31–47.

United States Food and Drug Administration (1987) Guideline on sterile drug products produced by aseptic processing. Food and Drug Administration, Maryland, USA.

*United States Pharmacopoeia XXIII* (1995) USP Convention, Rockville, Maryland.

# 4

# Enumeration of Micro-organisms

RONNIE MILLAR

*GlaxoWellcome Research and Development, Ware, Hertfordshire, SG12 0DP, United Kingdom*

## 4.1 Introduction

Microbiologists from all disciplines are concerned with estimating the number of microbial contaminants in samples under examination. This estimate in turn can be used to assess the risk presented to the recipient by that population of contaminants. Risk assessment forms an important part of the microbiologist's role.

Various techniques are available to the microbiologist seeking to count the number of micro-organisms in a given sample, the method of choice being determined by such considerations as cost, time-constraints, equipment required, sample specification, anticipated sample quality, the physical nature of the sample and, of course, personal preference. This chapter reviews the more commonly employed approaches to microbial enumeration that are generally available to the microbiologist working in pharmaceutical quality control. The practices described are 'traditional' procedures; recently developed 'rapid methods' are addressed only superficially here as the topic is more fully reviewed in Chapter 7.

## 4.2 Sample preparation

The first stage of the process is 'sample preparation', the purpose of which is to render the sample material into an homogeneous solution, suspension or emulsion that is suitable for analysis and determination of the microbial count. Depending on the nature of the sample, 'preparation' may involve heating, mechanical treatment, or possibly the addition of certain chemicals. In all cases the preparation should be such that the bioburden of the sample is not affected to give either an artificially high or low microbial count. This places considerable limitations on the preparation process. Incubation of the sample under conditions designed to revive dormant or damaged organisms [as described in *European Pharmacopoeia* (EP) 1997, section 2.6.13] is one such procedure which may lead to an artificially high count unless carefully controlled. An overview of preparatory procedures as commonly applied to different sample types is given below.

## 4.2.1  *Aqueous materials*

Materials that are aqueous in nature do not normally require any pre-treatment prior to enumeration. Very highly pigmented or antimicrobial materials may require dilution or possibly the addition of agents to neutralize antimicrobial properties before enumeration (see section 4.5).

## 4.2.2  *Soluble materials*

Other than the addition of sterile water, readily soluble materials do not require pre-treatment before enumeration. Some materials may require gentle heating or mechanical mixing to aid their solubility.

## 4.2.3  *Insoluble materials*

### *Non-fatty insolubles*

Here the objective is to prepare an homogeneous suspension of the material. This process may be aided by adding glass beads to the mixture of sample and diluent to assist in breaking up the sample. The mixture should then be shaken vigorously, either manually or aided by a vortex mixer or similar device. Repeated or continuous agitation may be necessary to prevent sedimentation during processing. The inclusion of a surfactant such as lecithin or polysorbate 80 in the diluent may also aid suspension of the material. The concentration to be employed may be determined empirically, but concentrations between 1–3% are likely to be satisfactory provided that foaming does not arise. It is necessary to validate this procedure in order to demonstrate the absence of surfactant toxicity for micro-organisms.

### *Fatty insolubles*

In order to prepare an homogeneous emulsion, mechanically assisted mixing, the addition of surfactant and use of heat are normally required to varying degrees, depending on the specific nature of the material. One approach is to add a small amount of surfactant (e.g. up to 5 g of polysorbate 80 to 10 g of sample) and glass beads to the sample. This mixture may then be placed in a shaking water bath or orbital shaker at not more that 40°C for the briefest time necessary to homogenize the material, not normally exceeding 20 min. The pre-heated diluent should be added and the mixture shaken to form an emulsion. Elevated temperatures should always be maintained for the minimum amount of time, as heat can adversely effect the microbial count, particularly if the sample microflora is sublethally damaged.

### *Gelatinous insolubles*

Again, heating and mechanically assisted homogenization of the material are normally required. Additionally, chilling of the material immediately prior to testing often assists sample preparation. A typical approach is to refrigerate the diluent at approximately 4°C for 1 h before adding the sample. The mixture of sample and diluent is then maintained at 4°C for 1 h, after which glass beads are added, and the sample mixture shaken vigorously and then held at approximately 35°C for 30 min, with regular shaking.

### 4.2.4  *Medical devices*

Here, the intention is to remove the microflora present on the surfaces of medical devices and to suspend it in a diluent; information on the procedures to be adopted is given in EN 1176 Part 2 (1996). The sample preparation stage normally requires a known number of devices to be added to a specified volume of sterile diluent. If a device is too large to be accommodated in the container(s) available, consideration must be given to alternative strategies of sampling a representative fraction of the device, or dismantling or subdividing it in such a way that it fits into standard laboratory vessels. The devices are then treated to suspend the microflora in the diluent. Options for removal of the bioburden from the device include: stomaching; ultrasonication; vortexing; flushing; disintegration (blending); swabbing; agar over-laying or contact plating. Care should be taken to ensure that the micro-organisms are effectively removed from the surfaces of the devices, but that the stresses involved do not adversely affect the microbial count. The addition of surfactants to the diluents can aid removal. Once suspension of the contaminating organisms has been achieved, the suspension is treated in the same way as any other aqueous sample.

### 4.2.5  *Pressurized products*

Sample materials contained within pressurized containers present a particularly stringent challenge to the microbiologist. The formulation often consists mainly of highly volatile propellant which is quickly lost when the container is breached, and is therefore difficult to test. The pressure and toxicity of the propellants additionally present safety concerns. At present, the best pharmacopoeial guidance [*United States Pharmacopoeia* (USP), 1995] for this type of product requires the container to be frozen in a dry ice/alcohol mixture, after which the container should be cut open. When the propellant has evaporated the remaining product should be tested. This is satisfactory provided that sufficient product remains behind for testing, though often only residues of product remain. In addition, the extremely low temperature exerts a considerable stress on any micro-organisms and introduces further safety hazards to the testing technician.

An alternative approach is to evacuate, at room temperature, the contents of the container through a sterile needle onto the surface of a liquid medium on which a viable count can be undertaken after the propellant has evaporated. Impaction on the liquid surface is preferable to bubbling the contents through the liquid because this may cause excessive frothing.

### 4.2.6  *Gases*

A similar approach may be employed when testing the bioburden of gaseous mater-ials, except that here the gas should be bubbled through a suitable diluent rather than impacting on to the liquid. This should be done for a specified period, after which the diluent should be tested as for aqueous samples.

### 4.2.7 *Antimicrobial materials*

Where the sample material has antimicrobial properties, these must be eliminated during the test procedure in order to permit any viable micro-organisms present to grow. This may be accomplished by filtration, where the sample is physically removed from the microflora by washing of the membrane. Alternatively, dilution also reduces or removes the antimicrobial effect that the sample may have; however, repeated dilution of the sample also serves to reduce the counting sensitivity of the test method. Where such procedures are ineffective or impractical, the addition of chemical inactivating agents should be considered (see Chapter 2, section 2.2.1; see also section 4.5). In some circumstances, all or a combination of these techniques may be required to neutralize the antimicrobial effect of the sample.

### 4.3 Counting methods

In contrast to their counterparts involved in the chemical analysis of pharmaceuticals, pharmaceutical microbiologists have been largely restricted to test methods that have changed little over the years, many of which would be quite familiar to Louis Pasteur himself. However, over the past 10 years or so, considerable advances in electronics, computer science and imaging technology have facilitated the development of a range of sophisticated automatic and 'rapid' techniques. This new generation of microbiological test methods is based on a relatively diverse range of biochemical principles. The methods rely on: microscopy to detect discrete microbial cells that have been selectively stained by an 'epi-fluorescent' stain to distinguish viable from non-viable cells; the detection of microbial ATP; detection and measurement of microbial growth by following conductance or impedance in the growth medium caused by microbial metabolites; and enzyme-linked immunosorbent assays (ELISA), polymerase chain reaction (PCR), etc. Many rely heavily on increasingly sophisticated computer technology, and consequently the hardware is relatively expensive. In addition to 'rapid' methods whose objective is to provide a measured viable count in a short time, the application of new technology to microbiology has also provided many devices to reduce the amount of manual intervention required to perform microbial counts. Such automated methods and rapid methods are described in Chapter 7. The remainder of this chapter is devoted to consideration of the relative merits of the more 'traditional' methods used in the pharmaceutical microbiology laboratory.

Based on more traditional technology, an array of microbial enumeration techniques is available, although several slide counting chamber-based methods for example are perhaps used more widely in food and clinical microbiology, but rarely in routine pharmaceutical microbiology. The methods most frequently used to test pharmaceutical raw materials and products are characterized by the time and resource required to perform them, equipment required, running costs, sensitivity of microbial detection, etc. Although it is difficult to identify published comparisons of methods in terms of accuracy, precision, etc. (partly because such comparisons will be

**Table 4.1**  Suitability of common methods according to sample type

| Physical nature of sample material | Appropriate enumeration technique | | | | |
|---|---|---|---|---|---|
| | Pour plate | Membrane filtration | Spread plate | Miles and Misra | Most probable number |
| Aqueous and water-soluble | + | + | (+) | (+) | (+) |
| Insoluble/non-fatty | (+) | – | (+) | – | (+) |
| Insoluble/fatty | (+) | – | + | – | + |
| Insoluble/gelatinous | (+) | – | + | – | – |
| Devices | – | + | (+) | (+) | (+) |

+, Recommended; (+), Suitable, not recommended unless specific circumstances preclude the use of other methods; –, Unsuitable.

influenced by the skill of the operator), it has long been established that different methods of counting may yield different recovery values (Cowan and Steiger, 1976). Additionally, certain methods are only suitable for certain types of sample. All of these factors must be considered when choosing a counting method; a general guidance on the suitability of commonly used 'traditional' methods according to sample type is provided in Table 4.1.

### 4.3.1  *Pour plating*

This is perhaps the most widely used method of enumeration in pharmaceutical microbiology quality control testing (Figure 4.1). A measured portion of the pre-pared sample is added to empty Petri dishes, typically two plates for each chosen agar type. Approximately 20 ml of molten agar, at a temperature not higher than 45°C, is then added to the Petri dishes. The plates are gently agitated to distribute the sample uniformly in the agar, the agar is allowed to set, and the plates are then incubated. After incubation, any micro-organisms present in the sample should have grown to form distinct colonies on and within the agar.

   This method is suitable for most forms of sample materials. It is a relatively simple and quick method to perform, being most efficient where there are a number of samples to test and the plates can then poured on the same occasion. Another advantage of the technique is that it only requires agar and Petri dishes, and so is relatively inexpensive to perform. One disadvantage is that where the preparative procedure involves dilution of the sample – which is often necessary if the sample material is microbiologically inhibitory – the dilution results in a relatively small amount of sample being tested. This produces a corresponding reduction in the test's sensitivity of microbial detection. For example, based on pharmaceutical recommen-dations, it is common practice to add 10 ml or 10 g of sample to 90 ml of diluent, giving a 1 : 10 dilution. If 1 ml of the preparation is added to each Petri dish, then only 0.1 ml or 0.1 g of sample is tested, giving a sensitivity of detection of 10 colony forming units (c.f.u.) per 1 ml or 1 g of sample. The volume of sample preparation that is added to the Petri dish is restricted by the fact that if the molten agar is diluted excessively it will not

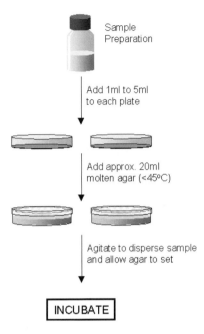

Sample
Preparation

Add 1ml to 5ml
to each plate

Add approx. 20ml
molten agar (<45°C)

Agitate to disperse sample
and allow agar to set

INCUBATE

**Figure 4.1**   The pour plate procedure.

set properly. Normally, 1–5 ml of sample preparation is added, though it is possible of course to use large diameter Petri dishes to accommodate larger volumes of sample.

A further limitation of this technique is that pour plates prepared from samples that are very poorly soluble may be difficult to read due to the presence of non-viable particles in the agar, which may mask viable colonies. Similarly, samples that have been prepared as emulsions may cause intense clouding of the agar and prevent visual detection of microbial colonies. In some cases the identification of viable particles in the agar may be assisted by the addition of a suitable staining agent such as triphenyltetrazolium chloride (TTC), which causes microbial colonies to stain red.

When preparing pour plates, the operator must always consider that excessively warm molten agar can exert a critical stress on the sample microflora and so artificially reduce the detected count. This is significant as the micro-organisms contained within many samples may already be sublethally injured and therefore sensitive to such a temperature shock. As a rule of thumb, the bottle of agar should be no hotter than would permit it being held comfortably in the hand.

### 4.3.2   *Membrane filtration*

Due to its simplicity and high sensitivity, membrane filtration is also a widely used test. It is only suitable for aqueous and soluble sample materials, as insoluble materials cause the membrane to become blocked. The technique is particularly useful with those samples that exhibit antimicrobial properties.

Typically, a measured volume of sample preparation, normally corresponding to at least 1 g or 1 ml of sample, is filtered under vacuum through a sterile membrane filter (Figure 4.2). The pore size of the membrane must not exceed 0.45 μm to be confident of

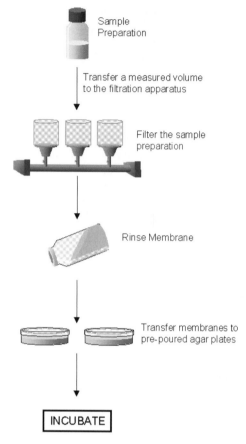

Sample Preparation

Transfer a measured volume to the filtration apparatus

Filter the sample preparation

Rinse Membrane

Transfer membranes to pre-poured agar plates

INCUBATE

**Figure 4.2**   Enumeration of organisms by membrane filtration.

retaining micro-organisms that might be present in the sample. The membrane is washed once or several times (depending on the nature of the sample), using diluent which may contain an added inactivation agent (see section 4.5). The membrane is then transferred aseptically to the surface of an agar plate (which may also contain an inactivation agent), and incubated. Any micro-organisms retained on the membrane will utilize nutrients from the agar and form distinct colonies on the membrane surface. It is therefore important that the membrane is carefully placed onto the agar, ensuring that there is good contact between membrane and agar. Care should be taken to avoid forming air bubbles between the membrane and agar as the membrane is positioned on the agar surface.

    Many forms of apparatus are available to assist the microbiologist wishing to use this technique; these are available from several manufacturers including Sartorius, Millipore and Gelman. The equipment ranges from multi-use precision-manufactured sterilizable steel and glass filtration units to simple disposable plastic pre-sterilized filtration units (Denyer and Hodges, 1999). With multi-use systems the filter funnel/membrane assemblies are often made of glass and then clamped to the membrane-supporting base unit, which is then attached to a stainless steel manifold, Buchner flask and pump or perhaps to a dedicated vacuum pump.

The sample contact parts of multi-use systems must be autoclaved before use and may be used many times. Such systems therefore require some effort to prepare and sterilize the equipment; however, the only disposable items are the membranes, diluent and agar plate. Disposable units are normally purchased pre-sterilized by irradiation. They are convenient, quick and easy to use, but may be relatively expensive if used continuously. Additionally, disposable filtration systems are available where the used membranes sit on a broth cartridge and are not transferred to agar plates, thus eliminating this opportunity for accidental in-test contamination and making the test simpler and quicker, and reducing the amount of disposable items.

As the sample passes through the membrane and is then further washed with a suitable volume of diluent, possibly containing antimicrobial inactivators, this technique is particularly applicable to aqueous or soluble materials that are anti-microbial by nature. Similarly, large volumes of sample may be tested using this technique, providing an extremely sensitive test. The technique is therefore well suited to materials that possess a very tight microbial specification. A disadvantage of the system is that the area of the membrane available to support microbial growth is relatively small, compared with, for example, a Petri dish. If the microbial count is unusually high it may prove to be uncountable due to confluent growth as colonies merge into each other.

The majority of samples may be filtered through cellulose nitrate filter membranes; however, alcoholic solutions affect the physical substance of such membranes, altering the effective pore size and often causing the membrane to blister or curl. This obviously has an adverse effect on the performance of the membrane and so it is recommended that alcohol-based samples only be filtered through cellulose acetate membranes.

### 4.3.3 *Spread plating*

This method is particularly suitable for insoluble materials that are fatty or gelati-nous in nature, and for the culturing of fungi.

Typically, between 0.1 ml and 0.3 ml of sample preparation is added to the surface of a pre-poured agar plate. The sample is spread over the whole surface of the plate with a sterile spreader, and the plate is then incubated. After incubation, any micro-organisms present in the sample should have grown to produce distinct colonies on the agar surface. The advantage of this technique is the ease with which it is performed, not requiring any agar to be pre-melted or any provision of sterile equipment with the exception of a sterile spreader (pre-irradiated disposable plastic or re-usable steriliz-able glass). It is a relatively low-cost technique, economical in its use of time resource, and avoids the potential risk of heat stress that poured molten agar may exert on the sample microflora. In addition, the colonies (of aerobes or facultative anaerobes) which arise on the agar surface tend to be larger than those within a pour plate since there are no effects of limited oxygen availability; this may facilitate the recognition and counting of some species. On the negative side, however, the sample is not mixed with the agar but is spread on top of it, thus the degree of dilution of antimicrobial substances by the agar itself is probably lower with a spread plate than a pour plate technique since dilution in a spread plate method relies on *diffusion* of the anti-microbial substance from the sample into the agar. Additionally, as the spreading of the sample on the agar surface may be irregular, an occasional observation with this technique is localized patches of uncountable confluent growth. The technique is therefore best suited to samples where the microbial count is likely to be low. A further weakness of this method is that it requires a relatively high amount of manual intervention and activity over the exposed plate in order to spread the sample thoroughly, which provides an opportunity for accidental contamination during testing.

### 4.3.4  *Miles and Misra plating*

Also known as the 'drop count' method, this is perhaps the simplest and least accurate test available to the microbiologist. It is best suited to aqueous or highly soluble samples where test sensitivity is not a priority.

A measured drop of sample is placed onto the surface of an agar plate. As with other methods this can be repeated with a series of dilutions of the sample preparation. When dry, the plate is incubated, after which any colonies present in the 'drop zone' are counted and the count corrected for the volume and dilution tested. The procedure is quick and easy to perform and requires less resource than any other enumeration procedure – indeed, several sample dilutions or replicate droplets may be tested on a single 9-cm Petri dish. The volume of sample plated using this technique is not always accurately measured, but simply dispensed from a standard '50 dropper'-type Pasteur pipette (approximately 0.02 ml). Reproducibility of the droplet size depends not only on the accuracy of the pipette, but also on technique. The pipette should be held vertically and the droplet dispensed not more than 2 cm from the agar surface in order to avoid splashing. Potential for variation in the plated volume may cause variation in results. Due to the small volume of sample tested, the sensitivity of the test is very low, being $500 \, \text{c.f.u.} \, \text{ml}^{-1}$ when the $1:10$ product dilution is plated. Also, since the sample covers a small agar surface area, only low numbers of colonies may be counted, normally not more than 30; consequently replicate plating of drops is recommended.

### 4.3.5  *Most probable number*

Although used widely in food and water microbiology, this enumeration technique is perhaps the least widely used in pharmaceutical microbiology. The statistical basis for the procedure is the assumption that micro-organisms are 'normally' distributed in liquid media. Replicate samples derived from the same source should theoretically vary in count, the variability being indirectly proportional to the microbial count, i.e. greater counts should give less variation between replicates. Replicate counts should yield a mean that is the *most probable number*. There are a number of variations of the test; the following (illustrated in Figure 4.3) is based on the procedures adopted by the United States (1995), British (1999) and European (1997) Pharmacopoeias. From a $1:10$ sample, serial dilutions of $1:100$ and $1:1000$ are prepared. From each dilution three separate 9 ml volumes of tryptone soya broth are inoculated with 1 ml. These sets of tubes will therefore contain 0.1 g or 0.1 ml, 0.01 g or 0.01 ml, and 0.001 g or 0.001 ml of sample material respectively. All broths should then be incubated at 30–35°C for at least 5 days.

After incubation, the broths should be inspected and the number of broths exhibiting growth recorded; where appropriate the broth should be streaked onto an agar plate and incubated to determine if growth is present in the broth. The most probable

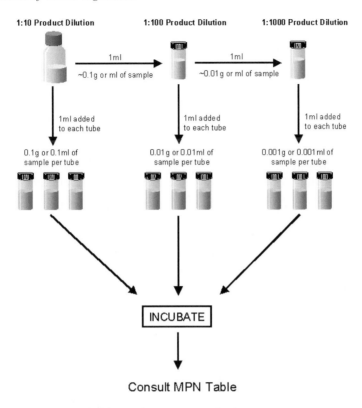

**Figure 4.3** The most probable number (MPN) method.

**Table 4.2** Interpretation table for the most probable number (MPN) method

| Number of tubes | | | Most probable number of organisms per g or ml of sample |
| --- | --- | --- | --- |
| 0.1 ml or 0.1 g sample per tube | 0.01 ml or 0.01 g sample per tube | 0.001 ml or 0.001 g sample per tube | |
| 3 | 3 | 3 | >1100 |
| 3 | 3 | 2 | 1100 |
| 3 | 3 | 1 | 500 |
| 3 | 3 | 0 | 200 |
| 3 | 2 | 3 | 290 |
| 3 | 2 | 2 | 210 |
| 3 | 2 | 1 | 150 |
| 3 | 2 | 0 | 90 |
| 3 | 1 | 3 | 160 |
| 3 | 1 | 2 | 120 |
| 3 | 1 | 1 | 70 |
| 3 | 1 | 0 | 40 |
| 3 | 0 | 3 | 95 |
| 3 | 0 | 2 | 60 |
| 3 | 0 | 1 | 40 |
| 3 | 0 | 0 | 23 |

number of viable micro-organisms contained in the sample material may then be determined by consulting Table 4.2. The technique is fundamentally qualitative, based on the detection of microbial growth by observing the presence or absence of broth turbidity/ gas in each tube. It is not dependent on the counting of discrete microbial colonies. In situations where the nature of the product causes the broth to become turbid or pigmented, microbial growth in the tubes may easily be confirmed by streaking the broth onto non-selective agar, such as tryptone soya agar and noting the presence or absence of growth. The technique may therefore be applied to almost any form of sample as insolubility of the sample does not hinder the reading of the test result.

The most probable number (MPN) method differs from other common methods by both the number of dilutions to be prepared and incubated, and by the degree to which it depends on statistics to convert the laboratory observations into the final numerical results related to the sample weight or volume. As a result, the MPN test is one of the least used enumeration techniques in pharmaceutical microbiology laboratories. Sensitivity of detection is an additional limitation of this procedure, the above test being able to detect no less than 23 c.f.u. per gram or millilitre of sample. Alternative versions of the test employ more tubes, and have a corresponding increase in sensitivity of detection.

## 4.4   Turbidimetric methods

The procedures described in section 4.3 are used to determine concentrations of micro-organisms when they are present at levels typical of acceptable bioburdens in pharmaceutical materials, i.e. not more than thousands per millilitre or gram. There are situations, however, where the pharmaceutical microbiologist needs to prepare suspensions of micro-organisms at a known concentration which are to be used in various pharmacopoeial methods, and the procedure here is usually to prepare a dense cell suspension, typically at a concentration of $10^6$ to $10^9 \, \text{ml}^{-1}$ and to dilute this appropriately. Situations in which standardized inocula are required include the following:

- In tests to evaluate the activity of biocides (for example, in pharmacopoeial preservative efficacy tests).

- For controls to demonstrate the suitability of a culture medium (for example, in sterility testing and detection of specified organisms).

- For controls to demonstrate the adequacy of a procedure to neutralize antimicrobial activity.

Bacterial suspensions typically become turbid at concentrations in excess of $10^7 \, \text{ml}^{-1}$, whereas suspensions of yeasts and moulds, because of the larger size of the individual cells, may be cloudy at concentrations which are 10-fold or even 100-fold lower. It is possible, therefore, to construct calibration plots which relate turbidity to cell concentration for all the common bacteria and yeasts used in pharmaceutical testing and for suspensions of spores (but not mycelia) of common moulds such as *Aspergillus* and *Penicillium* species. The data for such plots may be obtained from instruments designed specifically to measure light scattering (nephelometers) or light absorption (spectrophotometers or colorimeters). In addition, calibrated aqueous suspensions of such materials as barium sulphate or polystyrene in sealed glass tubes are used as reference standards, e.g. McFarlane tubes. The following points are worthy of note when constructing calibration plots:

1   The liquid used to suspend the microbial cells and the reference or 'blank' used to zero the instrument should be identical. Samples of broth autoclaved for different periods of time will darken to differing degrees, so a 'dark' broth read against a 'light' one will result in a positive instrument reading, even when there are no organisms in suspension. Also, samples of fresh broth will not have the same light-transmitting properties as used growth medium which suffers change due to microbial growth.

2   The turbidities of bacterial suspensions are higher when read on a spectrophotometer at low wavelengths. Thus, a wavelength of 430 nm might afford greater sensitivity than, say, 700 nm. The choice of wavelength is also determined by the absorption characteristics of the medium; in this respect, the yellow colour of many nutrient broths often requires a move to longer wavelengths.

3   Some bacteria produce coloured pigments which cause an artificially high instrument reading (which comprises the light scattering component and a light absorption component due to the pigment). The blue-green phenazine pigment of *Pseudomonas aeruginosa* for example absorbs significantly at low wavelengths, so a wavelength in excess of 470 nm is recommended for this organism.

4   The aggregation or 'clumping' of bacterial cells in suspension is a phenomenon which is influenced by their cultural conditions, and the recorded turbidity might be influenced by the degree of clumping.

5   The turbidity of a suspension depends on the surface area of the suspended cells. Bacilli in young cultures are usually longer, and thus possess a larger surface area, than cells in old (stationary phase) cultures; this might influence the slope of the calibration plot.

6   The correlation of turbidity and cell concentration is only linear up to an absorbance or optical density value of approximately 0.3. In cultures or suspensions which give readings higher than this, there is a departure from linearity because the bacterial cells in suspension on the side of the cuvette nearest to the instrument light source effectively cast a shadow over the bacteria more distant from the light, so the latter do not contribute fully to the instrument reading. The precise region of linearity in any particular suspending environment can be determined by experimentation.

Examples of turbidimetric calibration plots for the three bacteria recommended for pharmacopoeial preservative efficacy tests are provided in Chapter 10.

## 4.5   Culture media

The agar media employed in testing should be chosen with both the 'target' microbial population and the sample properties in mind. Tryptone soya agar (TSA, also called soya casein digest agar in the pharmacopoeias) is recommended for the general enumeration of bacteria. TSA plates should be incubated at 30 to 35°C for approximately 3 days. For the general enrichment of yeasts and moulds, Sabouraud dextrose agar is recommended by the pharmacopoeias. The formulation and low pH of this agar, the optional inclusion of antibiotics and incubation at 20 to 25°C are all designed to promote fungal growth and restrict bacterial growth. When there is a requirement to enumerate specific organisms, such non-selective media may be exchanged for selective media, chosen to suit the organisms in question. Information concerning preparation and storage of culture media, drying of plates, etc. is given in Chapter 2.

Diluents used for sample preparation and any subsequent product dilutions should have no effect on the microbial count during the time that they are in contact with the product. Examples of suitable diluents are physiological saline, phosphate-buffered (USP) or buffered peptone solution (BP, EP).

Depending on the nature of the sample, certain additives may be included in the diluent or agar used, or both if required. These may be surfactants such as polysorbate 20 or polysorbate 80, added simply to aid suspension or emulsion formation. These reagents however are frequently added because of their ability to neutralize the antimicrobial effect of certain sample materials, particularly quaternary ammonium compounds such as benzalkonium chloride. Lecithin and polysorbate 80 have indeed been employed in media for over 50 years for this purpose. A widely used combination of these materials is 0.3% soya lecithin (e.g. Azolectin$^{TM}$) and 2% polysorbate 80 (e.g. Tween 80$^{®}$). Other inactivators include penicillinase and *p*-aminobenzoic acid which may be employed to inactivate certain $\beta$-lactam and sulphonamide-type antibiotics, respectively (see Chapter 2, Table 2.1).

As mentioned earlier, the visualization of viable colonies can be significantly impaired by samples that are highly pigmented, insoluble, or which produce thick opaque emulsions. In such instances the addition of 1% TTC is recommended. This material is colourless in its normal oxidized state, but is reduced through microbial metabolism to produce insoluble formazan, a strongly red-coloured substance. This reduction is irreversible and so produces highly pigmented microbial colonies that are readily detected in an otherwise opaque/cloudy or particle-strewn agar plate or filter membrane.

## 4.6   Method validation

Although the testing laboratory is at liberty to select the specific test procedure for any given sample, and this may or may not be based on the guidance provided by the pharmacopoeias, regulatory authorities will always expect the laboratory to generate validation data in support of the application of the test. Even where the methods are tried, tested and widely established, as in this chapter, regulatory authorities require the testing laboratory to demonstrate that neither the sample material, test reagents nor any aspect of the test procedure, adversely affect the outcome of the test.

The pharmacopoeias all describe such a validation procedure, although they differ in detail. The validation normally consists of introducing a known low level of inoculum at a suitable point to ensure that the organisms are fully exposed to any microbial stresses introduced by the test procedure. The validation procedure is normally repeated using a collection of micro-organisms chosen to be generally representative of those encountered in the environment and during use. Microbial recovery in the absence and presence of sample is then measured and compared to assess the ability of the test procedure to recover organisms from the sample.

The procedure described in Figure 4.4 briefly outlines a validation procedure that is in keeping with the guidelines described in the current BP, EP and USP. This should be

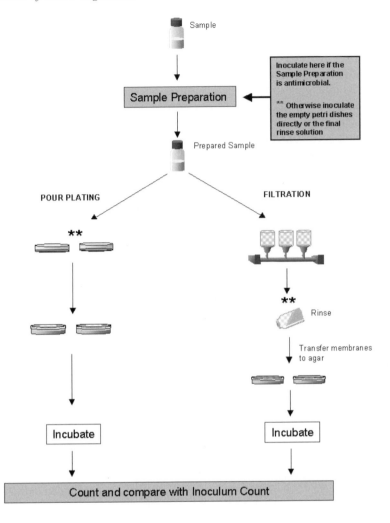

**Figure 4.4** Validation of counting methods.

performed separately for each of the following organisms: *Staphylococcus aureus*; *Escherichia coli*; *Bacillus subtilis*; *Candida albicans*; *Pseudomonas aeruginosa*; and *Salmonella abony* or similar. It is suggested that the inoculation level should be approximately 100 c.f.u. ml$^{-1}$ in the final cultured product preparation. With pour plates and membrane filtration techniques this should be not more than 150 c.f.u. per plate or membrane. Although the current EP (1997) indicates in section 2.6.12 that the count for a test organism should not differ by more than a factor of 10 from the calculated value for the inoculum (and this means that 10% recovery is acceptable), a more suitable minimum acceptable level of microbial recovery is in the range of 50% to 70% of the inoculum level. Microbial recovery should be determined as follows:

$$\text{Microbial recovery } (\%) = \frac{\text{Count in presence of sample}}{\text{Count in absence of product (inoculum)}} \times 100$$

In the case of medical devices, validation of bioburden determinations is complicated by the fact that it is necessary to demonstrate complete removal of the micro-organisms from the surfaces of the device (EN 1174-3, 1997). One method of valida-tion entails drying a standardized inoculum onto the surface of the device, but the drying process must not adversely affect viability. Quantifying the amount of inocu-lum introduced and ensuring its uniform distribution over the surface of the device may be difficult. One possible solution is total immersion of the device in a tared vessel containing the inoculum and recording the weight loss after the device has been removed and surplus liquid drained off. Clearly, if 100% recovery of the dried inoculum is achieved the procedure is satisfactory, but if the recovery is significantly less than 100% the question arises whether this is due to incomplete removal of the dried inoculum or to viability loss during the drying process itself. The use of bacterial or fungal spore suspensions will minimize this doubt since spores are usually resistant to desiccation.

## References

*British Pharmacopoeia* (1999) The Stationery Office, London.

COWAN, R.A. and STEIGER, R. (1976) Antimicrobial activity – a critical review of test methods of preservative efficiency. *Journal of the Society of Cosmetic Chemists*, **27**, 467–81.

DENYER, S.P. and HODGES, N.A. (1999) Filtration sterilization. In: RUSSELL, A.D., HUGO, W.B. and AYLIFFE, G.A.J. (eds), *Principles and Practice of Disinfection, Preservation and Sterilization*. 3rd edn. Blackwell, Oxford, pp. 733–66.

EN 1174 Part 2 (1996) *Estimation of the population of micro-organisms on products*. British Standards Institution, London.

EN 1174 Part 3 (1997) *Estimation of the population of micro-organisms on products; guide to the methods for validation of microbiological techniques*. British Standards Institution, London.

*European Pharmacopoeia* (1997) 3rd edn. EP Secretariat, Strasburg.

*United States Pharmacopoeia XXIII* (1995) United States Pharmacopoeial Convention, Rockville, MD.

# 5

# Identification of Micro-organisms

NORMAN HODGES

*School of Pharmacy and Biomolecular Sciences, University of Brighton, BN2 4GJ, United Kingdom*

## 5.1 Circumstances in which microbial identification is undertaken

Micro-organisms isolated from the pharmaceutical manufacturing environment, or which occur as contaminants of inadequately preserved products, require identification. There may also be other circumstances when process contaminants require rapid and reliable identification if heavy financial losses are to be avoided; contamination of a starter (inoculum) culture in a fermentation process is one such example. This chapter, however, is intended primarily for persons responsible for 'routine' identification of environmental and spoilage isolates rather than those working in more specialized fermentation or biotechnology facilities.

The need for identification is not, of course, simply to satisfy curiosity or to demonstrate a conscientious approach to control of the microbial quality of the manufacturing environment. Fast, accurate identification affords a practical benefit too, because certain organisms often arise in the same situations or from the same sources, for example, from dust, water or other raw materials, or from personnel. Thus, a reliable identification is often the first step in locating the source of a contaminant and formulating a strategy for its avoidance in future manufacturing batches.

## 5.2 Classes of organisms and identification strategies

Although there have certainly been occasions when organisms of significant pathogenic potential (e.g. *Salmonella* species, *Clostridium tetani*) have been isolated from pharmaceutical raw materials or manufactured medicines, these occasions have been relatively rare since the 1970s (Bloomfield, 2001; Spooner, 1996). The organisms most commonly found as contaminants of medicines and devices are not normally regarded as particularly hazardous, with *Bacillus* species and Gram-negative organisms such as *Pseudomonas*, *Enterobacter* and *Klebsiella* cited as the main groups and other Gram-negative species, staphylococci, *Candida*, *Aspergillus* and *Penicillium* species occurring less frequently (Watling and Leech, 1996). Since all of these organisms are classified by the Advisory Committee on Dangerous Pathogens (ACDP) (1995) in Hazard Group 1 or 2, it is evident that the identification of contaminants in pharmaceutical quality assurance laboratories can normally be undertaken in facilities conforming to ACDP Containment Level 2 (see Chapter 1).

The common contaminants of raw materials, finished medicines and devices are all bacteria, yeasts and moulds; viruses are not normally considered as contaminants except in vaccines and in certain specialized 'biotechnology' products. The strategies for identifying bacteria differ from those used for moulds because bacteria are not normally sufficiently different from each other in appearance to permit identification by microscopy. It is usually necessary to employ biochemical tests which examine the ability of the organism to produce particular enzymes; these tests are often incorporated in commercially available miniaturized test kits. Moulds, on the other hand, can normally be identified on the basis of microscopy once they have been induced to produce asexual spores (the appearance of which are often characteristic). Yeasts represent an intermediate situation in which microscopy is sometimes sufficient, but, more commonly, biochemical tests are required.

Most published identification schemes and commercially available kits are capable of identifying an isolate down to species level. In a number of situations this level of precision may not be necessary, and, provided that the pharmacopoeial 'specified micro-organisms' are not detected, it would be sufficient to classify contaminants simply as e.g. Gram-positive cocci, Gram-negative rods, bacterial spore-formers, yeasts or moulds; it is worth noting here that Gram-negative cocci are rarely encountered in this situation. This broad level of categorization can sometimes be achieved simply by observation of the cultural (macroscopical) and microscopical characteristics of the organism, and so the identification is completed more rapidly and cheaply. However, identification to genus level on the basis of macroscopical and microscopical characteristics is relatively rare, and for generic identification of bacteria it is usually necessary to employ commercially available biochemical tests kits, some of which are computer-driven or computer-assisted; these procedures are always necessary for identification to species level.

Rarely is it necessary in a pharmaceutical setting to distinguish strains within a species. If the need for strain differentiation does arise, it is normally undertaken by phage typing (examining the susceptibility of the isolate to bacterial viruses), by typing with bacteriocins (antimicrobial proteins secreted by certain bacteria), or by immunological methods. These procedures are best carried out by specialist laboratories and are outside the scope of this chapter.

## 5.3  Bacteria

### 5.3.1  *Cultural (macroscopic) characteristics*

The cultural characteristics and the macroscopic features of the organism growing on agar which may be used as an aid in the identification process are shown in Table 5.1.

It is rarely possible to determine the precise optimum growth temperature for an organism in a microbiology quality assurance laboratory, but some information may be gained on its preferred growth temperature simply by comparison of the rates of growth when cultures are incubated on routine non-selective media (nutrient agar, tryptone soya agar) at 35°C, 30–32°C and ambient temperature. If the culture grows

**Table 5.1**   Cultural and microscopical features of bacteria which may assist in their identification

| Cultural characteristics | Microscopical features |
| --- | --- |
| • Optimum growth temperature<br>• Ability to grow in anaerobic conditions<br>• Appearance of isolated colonies on agar: pigmentation, shape, size, surface markings, swarming, characteristic odour. | • Cell shape and size<br>• Gram-staining reaction<br>• Cell aggregation patterns, branching<br>• Motility<br>• Spore production |

best at temperatures lower than 35°C it is less likely to be a common human pathogen than if 35°C *is* the optimum; this fact may exclude several possible identifications. Similarly, a comparison of growth in air and in an anaerobic jar should indicate whether the organism is best described as a strict aerobe or a facultative anaerobe. Strict anaerobes are rarely recorded as contaminants in medicinal products, but this may, in part, simply be a reflection of a reluctance to undertake anaerobic bacteriology which is somewhat more demanding than the culture of species that grow in air. If anaerobic facilities are not available, simple observation of the growth characteristics of the organism in a tube of liquid medium may provide a valuable pointer to the identity. Strict aerobes, e.g. *Pseudomonas* species, many *Bacillus* species and many moulds, are likely to grow with a thick pellicle at the meniscus and little or no turbidity below the surface, whereas most Gram-negative enteric organisms such as *Klebsiella*, *Enterobacter* and *Proteus*, plus other Gram-negative rods such as *Serratia*, will show uniform turbidity throughout the tube.

Bacterial colonies growing on the surface of non-selective media may possess several distinguishing features. It is important to stress here that colonies should be well separated from each other because certain characteristic features may not be evident in areas of confluent growth; in other words, well-streaked plates and pure cultures are essential to obtain reliable information, whether this be from macro-scopical, microscopical or biochemical (test kit) methods. A procedure for streaking a plate to maximize the probability of obtaining isolated colonies is shown in Figure 5.1. A number of common contaminants produce pigments of characteristic colour on routine media. Many species of *Pseudomonas* for example produce green, blue or turquoise pigments, several *Serratia* species produce red pigments, and many skin cocci, e.g. *Staphylococcus* and *Micrococcus* species have colonies coloured yellow, gold or orange. Not only is the mere presence of the pigment characteristic but its diffu-sibility is another distinguishing feature. Thus, the phenazine pigments of *Pseudomonas* cross the bacterial cell membrane and diffuse throughout the agar so that the whole plate may become coloured; this is not the case with the other examples cited above in which only the colonies are pigmented.

The shape, size and surface markings of colonies are other characteristics which may facilitate identification. While the colonies of many enteric bacteria are approxi-mately circular, colourless and characterless, those of many *Bacillus* species are usually large (sometimes 2 cm or more in diameter) with characteristic surface markings, e.g. irregular or concentric circular ridges (Figure 5.2). Some colonial characteristics are almost unique to a particular genus; for example, the phenomenon known

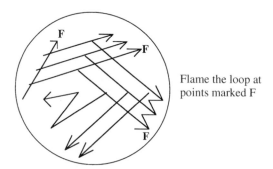

Flame the loop at
points marked F

**Figure 5.1**   A procedure for streaking a Petri dish which maximizes the probability of obtaining isolated colonies.

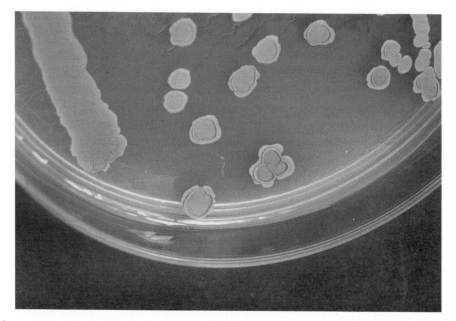

**Figure 5.2**   Colonies of *Bacillus megaterium,* showing the large diameter and surface markings (in this case concentric ridges) which are typical of many *Bacillus* species.

as 'swarming', which describes the prolific spreading of the colony as a thin film over the surface of the agar, is almost confined to *Proteus* species (Figure 5.3), although it may very rarely be seen with other vigorously motile organisms, e.g. vibrios.

### 5.3.2  *Microscopic characteristics*

Microscopic examination of a bacterial isolate will provide useful information on its possible identity. In order to obtain reliable information regarding the shape, size, motility, Gram-stain reaction and potential to exhibit filamentation, branching or spore production, it is necessary to have not only a good microscope with both bright-field and phase contrast operation, but also personnel who are adequately trained in its use. Unfortunately, because the basic operating principle of a

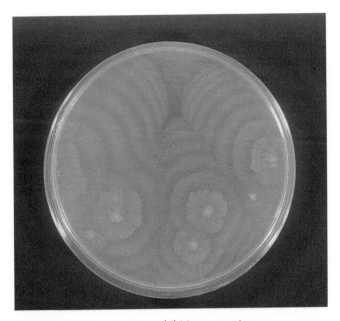

**Figure 5.3** Colonies of a *Proteus* species exhibiting swarming.

microscope is straightforward, training in microscopy is often minimal or non-existent. The consequence is that many laboratory staff use a microscope with little or no idea either how to illuminate the specimen properly or to obtain optimum phase contrast, and so fail to elicit all the information which is available.

Frequently, the most important item of microscopical information is the cell shape, i.e. rod or coccus (curved or spiral cells are rarely encountered as pharmaceutical contaminants). Many textbooks tend to convey the impression that these two shapes are mutually exclusive, without mentioning that very short rods and 'squashed' cocci also exist, and it is sometimes difficult to assign a shape with any degree of confidence. Cells which are easily recognizable as rods when actively growing and in liquid media can become very much shorter as they age, particularly when grown on agar. Here, there is no substitute for a good quality microscope nor an operator who can achieve the best possible resolution. The consequence of incorrectly assigning the cell shape is that an inappropriate identification scheme may then be pursued, with a significant waste of time.

Cell size may act as a pointer to the possible identity of an isolate in some cases. While the diameter of cocci tends to be rather similar at approximately 1–2 μm, the size of rod-shaped cells can vary substantially. Gram-negative bacteria tend, on the whole, to have cells of smaller dimensions than those of Gram-positive species, so most *Bacillus* species are undoubtedly larger than cells of enterobacteria for example. The extremes tend to be exemplified, on the one hand, by pseudomonads, which may be as small as $0.5 \times 1.0$ μm, e.g. *Brevundimonas* (formerly *Pseudomonas*) *diminuta* and, on the other by the large bacilli such as *Bacillus megaterium* at approximately $2.5 \times 10$ μm.

As with cell shape, the correct assignment of a bacterial isolate's Gram-staining reaction is of critical importance if time is not to be wasted. The Gram-staining procedure is described in many textbooks (Collins *et al.*, 1995; Singleton and

Salisbury, 1996), and will not be reiterated here; however, two points are worthy of note. Just as some textbooks may lead the reader to believe that the alternative cell shapes are mutually exclusive, some books do not give sufficient emphasis to the fact that the Gram-staining reaction may change with cell age or cultural conditions or be, simply, inconclusive. If the staining procedure is conducted correctly, Gram-negative species are unlikely to appear Gram-positive, but some Gram-positive organisms may appear negative in old cultures, or the colour of the stained cells may be intermediate between blue-purple and that of the counterstain (usually red). Gram staining, therefore, should be conducted on actively growing cultures wherever possible. The second point is that the use of authentic Gram-positive and Gram-negative controls stained on the slide either side of the specimen under investigation will improve the chances of recording the result correctly.

The manner in which cells aggregate together is usually a stable characteristic of a species. While it is certainly true that many bacteria show no regular geometric pattern of aggregation, e.g. *Staphylococcus aureus* cells exist in random clusters which are often described as similar in appearance to bunches of grapes, the cells of others regularly appear in chains, e.g. many *Bacillus* species and streptococci, tetrads (groups of four) e.g. *Micrococcus* species, or palisades in which bacilli tend to align themselves with their long axes parallel (e.g. corynebacteria). Pairs of cells (diplococci) are often cited in textbooks as yet another aggregation pattern, but the relevant organisms (meningococci, gonococci and *Streptococcus pneumoniae*) are not encountered as pharmaceutical contaminants.

Branching of hyphae to form a mycelium is a well-recognized characteristic of fungi, but it should not be overlooked that some true bacteria also exhibit branching. *Streptomyces* species, for example, form relatively stable (non-fragmenting) mycelia

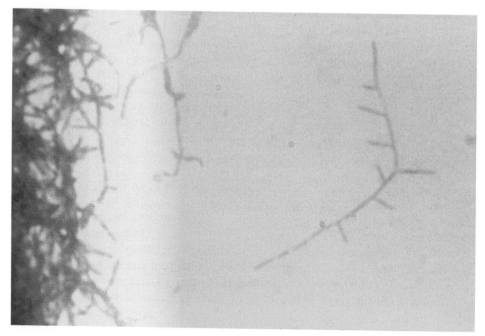

**Figure 5.4** Microscopical appearance of *Streptomyces mediterranei*, exhibiting the branching frequently seen in *Streptomyces*.

(Figure 5.4), and these organisms commonly occur in the soil and so may contaminate raw materials of vegetable origin or even mined minerals such as kaolin. A filament is an unusually elongated bacterial cell in which the length exceeds the width by approximately ten times or more. Species which exhibit branching, therefore, usually also exhibit filamentation, but elongated cell forms may also arise as a consequence of the cultivation conditions, particularly if antimicrobial agents are contained in a selective medium. Nalidixic acid in *Pseudomonas*-selective agar for example may increase the length of the cells dramatically so that they bear little resemblance to the standard description of this species; thus, care is required when interpreting the finding of filamentous bacteria in such circumstances.

The assignment of an isolate as motile or not is a relatively straightforward operation. Methods of confirming motility using semi-solid media and Craigie's tubes are described in some textbooks, but they are time-consuming and usually unnecessary if a good quality phase-contrast microscope is available. A drop of a young actively growing liquid culture is placed on an ordinary glass slide and a coverslip placed on top. Special 'hanging drop' slides which have a depression in one surface used to be employed, but are not now in common use. The organism is recorded as motile if individual cells can be seen moving simultaneously in different directions. Cells which merely oscillate about a fixed point are probably exhibiting Brownian motion rather than true motility, particularly if they are very small cells, and cells all moving simultaneously in the same direction are probably passively drifting in response to convection currents or evaporation of liquid from the edge of the coverslip. It should be recognized that cells which have the potential to exhibit motility may not do so if they are:

- old (stationary phase cultures);
- cold (ambient room temperature is usually 15°C below their incubation temperature, so metabolism is slowed accordingly, but this effect is unlikely to suppress motility entirely in cultures examined within a few minutes of removal from 37°C);
- stuck to the glass of the slide or coverslip (electrostatic or other interactions may cause immobilization);
- in a medium of high osmolarity (evaporation of water from the edge of the coverslip increases solute concentration); or
- in a medium containing an antimicrobial agent.

The presence of bacterial spores (endospores) in a culture represents an almost certain identification as a species of *Bacillus* or *Clostridium*, and of these the former is much more likely. There *are* other spore-forming bacteria; *Bergey's Manual of Determinative Bacteriology* (Holt *et al.*, 1994) describes 10 genera, but eight of these arise in specialized habitats and even clostridia, being anaerobes, are not very likely to be encountered as pharmaceutical contaminants. Spores appear under phase-contrast microscopy as bright spherical or oval structures which develop inside a rod-shaped vegetative cell (Figure 5.5). Their maturation is accompanied by the release of lytic enzymes so the cell in which they develop disintegrates and the spore is liberated, and in cultures showing extensive sporulation it may even be difficult to see an intact vegetative cell. The only significant problem in distinguishing spores is that several rod-shaped organisms (including *Bacillus* species) may accumulate spherical

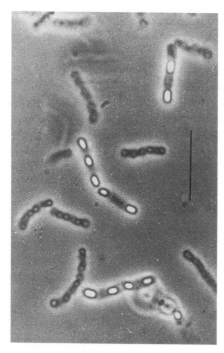

**Figure 5.5** Phase-contrast appearance of *Bacillus megaterium*, showing bright spores developed within the dark vegetative cells.

intracellular granules of poly-$\beta$-hydroxybutyric acid as a reserve food material in carbohydrate-rich media, and this appears bright under phase-contrast and may be mistaken for spores. The shape of the spore (spherical or cylindrical) and its position in the cell (central or terminal) may be influenced by cultivation conditions so they are of little diagnostic value. Whether or not the spore causes swelling of the sporangium (the vegetative cell in which the spore is formed) is a more stable, and so a more useful, distinguishing feature. Cell distension is much more commonly seen among clostridia than *Bacillus* species.

### 5.3.3 *Biochemical tests*

While it is certainly true that macroscopical and microscopical examination of a bacterial isolate may provide valuable pointers to its identity, it is also the case that many bacteria have few, if any, distinguishing features other than cell shape and Gram-staining reaction. These bacteria, therefore, are identified principally on the basis of their biochemical reactions, i.e. the results of tests to determine the ability of the organism to utilize selected substrates or produce characteristic end-products. Fermentation of sugars and other simple carbohydrates for example can be detected by the production of organic acids which alter the colour of pH indicators incorporated in culture media, and this may (or may not) be associated with the production of carbon dioxide as an ancillary product. Thus, the pattern of acid and gas production from selected 'sugars' may, in itself, be sufficiently characteristic to

**Table 5.2**  Some examples of biochemical tests used in schemes for bacterial identification

| Test category | Examples | Detected by |
|---|---|---|
| 'Sugar' fermentation | Lactose, xylose (and other molecules which are not true sugars e.g. mannitol) | Colour change of pH indicator due to acid fermentation products with, or without carbon dioxide |
| Pattern of glucose metabolism | Methyl red test | Acid detected by methyl red |
|  | Voges–Proskauer test | Ketones detected colorimetrically |
| Utilization of substrates | Citrate utilization | Indicator displays pH rise due to citric acid utilization |
| Proteolytic enzymes | Gelatin liquefaction | Liquefaction of gelatin-based gel |
|  | Milk agar | Casein-derived opacity reduced in agar surrounding bacterial growth |
| Production of characteristic enzymes or metabolic products | Oxidase test | Reduction of a redox dye to give a purple colour |
|  | Indole test | Colorimetric detection of indole as a metabolite of tryptophan |
|  | Nitrate reduction | Colorimetric detection of nitrite resulting from reduction of nitrate |
|  | Hydrogen sulphide production | Precipitation of black ferrous sulphide after interaction of $H_2S$ with iron salts in the medium |

provide a conclusive identification. More commonly, however, sugar fermentation tests are combined with tests to detect other metabolic activities (Table 5.2), and these may be performed by traditional methods using test tubes, Petri dishes and agar slopes, or they may be conducted in a miniaturized manner using microtitre plates, or by means of commercially available test kits which comprise fixed combinations of tests.

Table 5.2 is not intended to be comprehensive, but describes some of the principles upon which common biochemical tests are based. There are some 20–30 common biochemical tests which can be employed in identification schemes, and for certain of these modifications such as the use of synthetic, chromogenic substrates have been employed, e.g. *ortho*-nitrophenyl-galactoside as a substitute for lactose. These tests are comprehensively described elsewhere, and the reader should consult such texts as Cruickshank *et al.* (1975), Cowan and Steel (1993) or Collins *et al.* (1995) for details.

The outcome of these biochemical tests can be influenced by a variety of factors which need to be carefully controlled in order to achieve reliable results:

- Culture purity: the requirement for pure cultures is even greater here than in microscopy because test incubation times are sufficiently long for even a small number of contaminating cells to grow and give an erroneous result. A procedure for streaking a Petri dish which maximizes the chances of obtaining isolated colonies is shown in Figure 5.1.

- Inoculum size: very small inocula can result in false-negative results because the concentration of cells or enzyme achieved within the specified incubation period may be too low to give a positive result. Conversely, a high inoculum may produce a false positive due to carry-over of nutrients; this may occur in sugar fermentation tests for example.

- Incubation conditions: the duration and temperature of incubation specified in the test protocol are important, not only because inadequate incubation may give a false negative, but because in some situations the reaction responsible for a positive result may be reversed on extended incubation (see Chapter 6, Plate 22).

The routine use of positive and negative control cultures will clarify any doubts about the interpretation of a test result, and will be particularly valuable when pre-cise standard incubation conditions cannot be achieved, e.g. over weekend incuba-tion. Suitable control cultures are listed in the manuals of the major laboratory culture media suppliers, e.g. Oxoid, Difco and BBL and by Collins *et al.* (1995). When using biochemical tests on an *ad hoc* basis rather than as part of a commer-cially available test kit, it is also important to have information about the propor-tions of strains within a species which give a particular result. A significant proportion of atypical strains may not conform to the standard textbook result; for example, most strains of *Citrobacter freundii* are capable both of fermenting lactose and utilizing citrate as the sole source of carbon, but about 20% of isolates are negative for each of these tests. Bergey's Manual (Holt *et al.*, 1994) provides information on the proportion of strains which give positive results for many of the commonly encountered tests and organisms.

Conducting biochemical tests using traditional procedures (tubes, plates, slopes) rather than identification kits allows the operator to select the tests employed and to use as few or as many as necessary. In some cases this may result in cost savings, particularly if there are already strong pointers to an organism's identity and only a single test or a small number are required to provide confirmation. However, to use traditional methods on a routine basis there must be available:

- a relevant and broadly based identification scheme suitable for the organisms which typically arise in pharmaceutical contaminants;

- experienced operators who can select the most appropriate supplementary tests to be used for problem organisms and interpret the results; and

- a wide range of culture media stocks and adequate cold storage facilities for prepared media.

Generally, these disadvantages are considered to outweigh the advantages, and commercially available test kits are commonly used in preference. The API kits produced by BioMerieux, the market leaders, are well known, and these cover all the major groups of bacteria and yeasts which may arise as contaminants of pharmaceutical materials. There are, however, many other kits available from a number of manufacturers, and they – together with the API kits – are designed to identify organisms derived from broad classification categories, e.g. anaerobes or enteric bacteria, down to particular genera, e.g. staphylococci or streptococci or even individual species, e.g. *Candida albicans*: ten of the kits for identifying Gram-negative aerobic bacilli have been compared by Bennett and Joynson (1986). The API kits and some others, e.g. Enterotubes (Roche), consist of a fixed collection of tests which give patterns of results that are translated into a numerical value which, in turn, is converted into a species identification by reference to a code book. This affords the advantage that the user is also provided with an indication of the reliability of the identification – which is particularly valuable when the results pattern could arise from two or more different organisms. Perhaps the major disadvantage of this type of kit is that there are occasional problems in assigning a test result as positive or negative, because this involves a subjective assessment of the significance of a marginal colour change.

Most identification kits were designed originally for clinical specimens and occasionally problems arise with organisms of industrial importance or environmental isolates. However, manufacturers can often provide additional information when such organisms are encountered. Amy *et al.* (1992) have compared the suitability of different identification systems when applied to bacteria isolated from water. Operator input and scope for operator errors are reduced further by the incorporation of the identification kit principles into an automated system (e.g. Vitek) where there is little or no scope for interpretation by the operator and the results are read automatically and referred to the database in the computer which controls the instrument (see Chapter 7 for further details).

## 5.4  Fungi

Fungi (moulds and yeasts) are generally larger organisms than bacteria and they are structurally more complex and varied. As a consequence, identification, particularly that of moulds, is more heavily based upon a recognition of characteristic structural features (with a correspondingly reduced reliance on biochemical characteristics) than is the case with bacteria. Fungi are saprophytes, i.e. they obtain their nutrients primarily from dead and decaying organic matter, so pharmaceutical raw materials of vegetable origin usually have a significant fungal bioburden. Furthermore, most fungi are capable of producing spores which are resistant to drying and commonly exist as, or attached to, dust particles. Consequently moulds may comprise a substantial fraction of the atmosphere-derived product contaminants.

Fungi are eukaryotic, i.e. they are capable of sexual reproduction to give offspring which possess new combinations of genes and so differ from the parent organism; in this they differ from bacteria which can only reproduce asexually to give offspring identical to the parents. Since many fungi can produce both sexual and asexual spores at different stages of their life cycle and under different cultural conditions, and because the two spore types are borne on dissimilar and characteristic structures, the microscopic appearance of a mould may change significantly during the course of incubation. This means that identifying a mould on the basis of microscopy is a skilled operation requiring experienced personnel with access to good reference books containing detailed descriptions, diagrams and photomicrographs. The situation is further complicated by the fact that there are many different fungi which might realistically arise as contaminants of pharmaceutical materials yet most of these have little or no pathogenic potential; however, most of the reference books on fungal identification are largely or exclusively concerned with pathogens (Frey *et al.*, 1979; Koneman and Roberts, 1985; Larone, 1995). For these reasons some quality assurance laboratories which lack the resources for mycology may find it convenient for identification of fungi to be undertaken by specialist laboratories which offer a cost-effective service (e.g. CABI Bioscience, Egham, Surrey, UK).

### 5.4.1 *Yeasts*

The distinction between yeasts and moulds is not absolute. The term 'yeast' is generally taken to mean a unicellular fungus in which the individual cells are spherical or ellipsoidal and which commonly (but not invariably) reproduces by budding – a process in which a daughter cell arises from the parent as a localized outgrowth or bud. In a rapidly growing culture it is also common to see a secondary bud developing on a primary one, and the size and shape of the various cells in such a culture may vary significantly, i.e. the cells are described as pleomorphic (of variable appearance). A mould, on the other hand, usually consists of a tangled mass (mycelium) of filaments or threads (hyphae), some of which are specialized for absorption of water and nutrients (vegetative mycelium) or for reproduction (aerial mycelium). However, some organisms which are described as dimorphic can exist in either form depending upon the cultural conditions. Thus, *Candida albicans* for example often appears to have the microscopic appearance of a typical budding yeast, but on cornmeal agar it exhibits pseudohyphae in which the buds are elongated, and two buds may arise side by side at an angle of 45° to each other, thus conferring a branched appearance which resembles the mycelium of a conventional mould.

Yeasts growing on solid media often produce large colonies which look similar to those of bacteria, but yeasts are more frequently shiny and mucoid and quite often pigmented, with yellow, orange and red colonies being common. When examined under the microscope many yeasts do not have sufficient distinguishing features to permit their identification by this means alone, and biochemical tests are normally employed. These tests which largely determine patterns of fermentation and assimilation of various sugars were described by Collins *et al.* (1995), and by Koneman and Roberts (1985) who also described a more comprehensive yeast identification scheme. Again, there are API and other biochemical test kits available for yeasts, but the problem of a database biased towards pathogenic species remains.

### 5.4.2 *Moulds*

Biochemical tests play little or no part in the identification of moulds which are distinguished largely on the basis of their cultural characteristics and microscopical appearance. Mould identification, however, cannot be divorced from the subject of their classification, and it is important to recognize that this is even less straightforward than bacterial classification. A major confusing factor is that the asexually reproducing and sexually reproducing forms of the same organism may bear different names at both genus and species level, and this can result in common genera such as *Aspergillus* and *Candida* arising in two different classes, orders or families. As the possession of a sexual stage in the life cycle is a principal determinant of fungal classification and no such stage has been identified for many organisms, there exists a major category called deuteromycetes (also known as fungi imperfecti) to accommodate such organisms. Fortunately, almost all organisms of pharmaceutical interest fall either into this (large) group or into the zygomycetes, so the many other fungal classes can be disregarded in this context. Identification down to genus level for a substantial proportion of common contaminants can be undertaken in a pharmaceutical microbiology laboratory, provided that the operator has an adequate understanding of the descriptive terminology used in the standard reference books. While a detailed identification scheme is outside the scope of this chapter, the major distinguishing features of common moulds, together with explanations of the terminology, are considered below.

As with bacteria, growth rate (increase of colony size), pigmentation and surface markings are the most obvious features which can be used as pointers to the identity of a mould (Table 5.3). A temperature of 25°C is frequently used for the incubation of mould isolates, and the majority of organisms which arise as common contaminants produce visible colonies within 2–5 days at this temperature; infrequently a substantially longer period, even up to 30 days may be required. The fastest growing moulds are the zygomycetes which includes organisms such as *Mucor* and *Rhizopus* species. These frequently cover the surface of a standard Petri dish in 2 or 3 days incubation, and may then go on to fill all the available space above the agar and even lift the lid; deuteromycetes such as *Aspergillus* and *Penicillium* species are more likely to give colonies of a few centimetres diameter after 5 days.

Pigmentation of mould colonies may arise either because the hyphae themselves are pigmented, or because the spores are. If the hyphae themselves are pigmented (usually brown or black), the organism is described as dematiaceous; the term hyaline is used to describe organisms with unpigmented hyphae. If the colour of the colony is due to pigmentation of spores (as with *Aspergillus niger*, for example), this can usually be seen using a hand lens or with the naked eye, but even in cases of doubt another clear means of recognition is a colony having a white margin (which is young and still growing) with a coloured centre (which is mature, non-growing and containing developed spores; Figure 5.6); furthermore, a dematiaceous mould is black on both surfaces, but

**Table 5.3** Some macroscopical and microscopical characteristics used to distinguish moulds

| Cultural and macroscopical features | Microscopical features |
| --- | --- |
| • Growth rate: 2–30 days before appearance of colonies | • Hyphal width: from 1 to $\geq 50\,\mu m$ diameter |
| • Hyphal pigmentation: hyphae (as distinct from the spores) may be pigmented | • Hyphal branching: at right angles, dichotomous (45° or Y-shaped) or irregular |
| • Colony pigmentation: pigmentation associated with spore development in the colony centre | • Hyphal septa: septa (cross walls) present or absent |
| • Surface ridges (or rugae): ridges or folds radiating from centre of the colony | • Spore development: asexual or sexual or both (not simultaneously) |

**Figure 5.6** A culture of *Aspergillus niger* in which the central area of the colony shows black pigmentation due to spore production.

the underside of an *Aspergillus niger* colony would not be black. The possession of surface or underside ridges (also called rugae) is another feature sufficiently characteristic of certain species to act as a useful clue to the identity of an isolate, and this too is often exemplified by aspergilli (Figure 5.7).

Moulds are usually examined microscopically by directly removing a small fragment of the mycelium using a needle, and mounting it under a coverslip in a stain such as lactophenol–cotton blue. This stain serves both to impart colour and to reduce the risk of laboratory-acquired infection because the phenol component is fungicidal. Selecting an appropriate part of the colony for examination is important because the most useful diagnostic features are the spores and the structures upon which they are formed. If the specimen to be examined is taken from the very edge of the colony, it may contain no spores because this region is still actively growing and too immature; if, on the other hand, it is taken from the centre, the characteristic structures on which the spores are formed may have partially or totally disintegrated. For a mould which produces pigmented spores, therefore, it is normal to remove the sample from just within the pigmented region. With a colourless colony the sample is usually taken either from a point approximately one-third of the way from the periphery to the centre or from any raised central area which may be indicative of spore development.

Under the microscope the hyphae of common moulds differ most obviously in terms of their diameter, their degree of branching and possession of septa (cross walls). Hyphae may vary in diameter from as little as 1 μm to 50 μm or more, and the zygomycetes usually possess hyphae which are broader (6–25 μm) than those of the deuteromycetes. Hyphal length is not normally a particularly useful diagnostic character because it varies with age and maturity of the colony. Some cultures

**Figure 5.7**   The underside of the colony illustrated in Figure 5.6. The lines radiating
from the centre to the periphery (rugae) are a characteristic feature of *Aspergillus niger.*

show a tendency to hyphal fragmentation with increasing age and hyphal fragments
are often viable, so they, as well as spores, may initiate spoilage when introduced
into vulnerable pharmaceutical products. In some moulds the hyphae may be
unbranched, but in those species where branching arises, its form may be characteristic, e.g. the branch may regularly arise at 90° or 45° to the primary hyphae. However,
the possession of septa is a more important distinguishing microscopic feature. The
zygomycetes typically form a branched non-septate mycelium, whereas deuteromycetes give rise to a septate mycelium. If there are no septa, the hyphae may contain
multiple nuclei at intervals along their length (described as coenocytic), although in
some moulds the septa arise relatively infrequently. As septa are such a fundamental
determinant in identification schemes, it is important that this characteristic is
correctly recorded. Incorrectly assigning the possession of septa in moulds can be
both misleading and time-wasting.

Although a very small proportion of moulds isolated from pharmaceutical
materials may be capable of forming sexual spores, such spores are rarely seen,
and by far the most important distinguishing feature is the asexual spores. The two
major groups of common mould contaminants, i.e. zygomycetes and deuteromycetes
develop spores on markedly different structures and identification is based upon the
appearance of these structures. A detailed description of the asexual spore-bearing
apparatus of the majority of common moulds is outside the scope of this chapter,
and excellent diagrams and photomicrographs are to be found in a variety of books
(Frey *et al.*, 1979; Koneman and Roberts, 1985; Larone, 1995). It is essential to
understand the descriptive terminology before use can be made of these reference
sources and Figure 5.8, which illustrates an example of a species from each major
group, may be of value in this respect. It should be emphasized, however, that for both

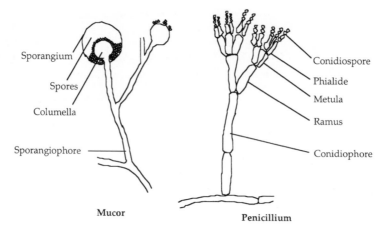

Sporangium

Spores

Columella

Sporangiophore

Conidiospore

Phialide

Metula

Ramus

Conidiophore

**Mucor**

**Penicillium**

**Figure 5.8** Diagrammatic representation of the asexual spore-bearing structures of *Mucor* and *Penicillium*.

the zygomycetes and deuteromycetes many species variations of these typical structures may exist and there are other specialized descriptive terms in use, many of which are defined by Singleton and Salisbury (1996).

The vegetative mycelium of zygomycetes is non-septate, branched, and grows more or less horizontally over the nutrient substrate which may be penetrated by hyphae in a manner analogous to the roots of green plants penetrating soil. During asexual sporulation the spores develop within a sac-like structure called a sporangium which, in turn, arises at the end of an aerial hypha (sporangiophore) growing perpendicular to the substrate. The sporangium ruptures on maturity and the numerous spores contained inside are released into the atmosphere. A further structure called a columella might be evident, and this may be merely an extension of the sporangiophore into the sporangium or a more prominent spherical or conical structure extending from the base of the sporangium.

The asexual spores of deuteromycetes are termed conidia, and these differ fundamentally from sporangiospores in that they do not develop within a membrane. In genera such as *Penicillium* and *Aspergillus* a conidiophore arises perpendicular to the substrate, and this is directly analogous to the sporangiophore of *Mucor* and *Rhizopus*. The conidiophore of penicillia, however, may branch two or three times to give structures termed, successively, rami, metulae and phialidae, and the conidiospores are formed by budding at the ends of the last of these structures.

## References

Advisory Committee on Dangerous Pathogens (1995) *Categorisation of Biological Agents According to Hazard and Categories of Containment*, 4th edn. HMSO, London.

AMY, P.S., HALDEMAN, D.L., RINGELBERG, D., HALL, D.H. and RUSSELL, C. (1992) Comparison of identification systems for classification of bacteria isolated from water and endolithic habitats within the deep subsurface. *Applied and Environmental Microbiology*, **58**, 3367–73.

BENNETT, C.H.N. and JOYNSON, D.H.M. (1986) Kit systems for identifying Gram-negative aerobic bacilli: report of the Welsh Standing Specialist Advisory Working Party in Microbiology. *Journal of Clinical Pathology*, **39**, 666–71.

BLOOMFIELD, S.F. (2001) Microbial contamination: spoilage and hazard. In: DENYER, S.P. and BAIRD, R.M. (eds), *Guide to Microbiological Control in Pharmaceuticals*. 2nd edn. Ellis Horwood, Chichester, In press.

COLLINS, C.H., LYNE, P.M. and GRANGE, J.M. (1995) *Microbiological Methods*, 7th edn. Butterworth-Heinemann, London.

COWAN, C.T. and STEEL, K.J. (1993) *Manual for the Identification of Medical Bacteria*, 3rd edn. Cambridge University Press, Cambridge.

CRUICKSHANK, R., DUGUID, J.P., MARMION, B.P. and SWAIN, R.H.A. (1975) *Medical Microbiology*, Volume 2, 12th edn. Churchill Livingstone, London, pp. 170–89.

FREY, D., OLDFIELD, R.J. and BRIDGER, R.C. (1979) *A Colour Atlas of Pathogenic Fungi*. Wolfe Medical Publications, London.

HOLT, J.G., KRIEG, N.R., SNEATH, P.H.A., STALEY, J.T. and WILLIAMS, S.T. (1994) *Bergey's Manual of Determinative Bacteriology*, 9th edn. Williams & Wilkins, Baltimore, pp. 559–61.

KONEMAN, E.W. and ROBERTS, G.D. (1985) *Practical Laboratory Mycology*, 3rd edn. Williams & Wilkins, Baltimore.

LARONE, D.H. (1995) *Medically Important Fungi*, 3rd edn. ASM Press, Washington DC.

SINGLETON, P. and SALISBURY, D. (1996) *Dictionary of Microbiology and Molecular Biology*, 2nd edn. Wiley, Chichester.

SPOONER, D.F. (1996) Hazards associated with the microbiological contamination of cosmetics, toiletries and non-sterile pharmaceuticals. In: BAIRD, R.M. and BLOOMFIELD, S.F. (eds), *Microbial Quality Assurance in Cosmetics, Toiletries and Non-sterile Pharmaceuticals*, 2nd edn. Taylor & Francis, London, pp. 9–30.

WATLING, E.M. and LEECH, R. (1996) New methodology for microbiological quality assurance. In: BAIRD, R.M. and BLOOMFIELD, S.F. (eds), *Microbial Quality Assurance in Cosmetics, Toiletries and Non-sterile Pharmaceuticals*, 2nd edn. Taylor & Francis, London, pp. 217–34.

# 6

# Pharmacopoeial Methods for the Detection of Specified Micro-organisms

NORMAN HODGES

*School of Pharmacy and Biomolecular Sciences, University of Brighton, BN2 4GJ, United Kingdom*

## 6.1 Introduction and scope

Tests for the detection of four named bacteria (*Escherichia coli*, salmonellae, *Pseudomonas aeruginosa* and *Staphylococcus aureus*) in pharmaceutical raw materials and finished products are described in the *United States Pharmacopoeia* (USP) (1995), and in the *European Pharmacopoeia* (EP) (1997) and *British Pharmacopoeia* (BP) (1999) (where the tests are identical). The EP and BP additionally describe a test for the detection of *Clostridium perfringens*. These organisms are singled out for special attention in this way because they represent particular infection hazards to the patient, or because their presence is a criterion of the quality of raw materials.

The detection tests are typically applied to raw materials of natural or biological origin (starches, gums, gelatin, talc, etc.) where there may be a background of unrelated organisms which may grossly outnumber the species of interest. Consequently, specific procedures for sample preparation and preliminary enrichment cultures are used to permit the organism of interest to increase in concentration and so become more readily detectable. The final stages of the detection tests invariably involve the use of solid selective media which, after incubation, are examined for the presence of colonies conforming to the standard textbook descriptions of the bacteria in question; any such colonies are then subjected to confirmatory biochemical or immunological tests. It is these final stages of the testing procedures which are the most problematical, because the textbook (or pharmacopoeial) descriptions of the bacteria are often imprecise and several relatively harmless organisms can also conform to these descriptions – leading to the possibility of false-positive identifications. The problem is compounded by the paradox that the tests are often undertaken by personnel working for companies using high-quality raw materials in which the organisms are rarely, if ever, encountered.

Reliable test results are dependent upon correct visual recognition of colonies; the old cliché that a picture is worth a thousand words is undoubtedly true in this case. This chapter, therefore, is intended to explain the principles and limitations of the tests, but it is primarily a comprehensive photographic account of the final stages of the testing procedures showing the typical appearance of the test bacteria on the EP- and USP-recommended media and their reactions in the various confirmatory biochemical tests. Corresponding pictures are included of other bacteria which are similar to, and frequently confused with, the bacteria that are the subject of the pharmacopoeial tests.

## 6.2   Significance and applicability of microbial limit tests

The microbiological quality of non-sterile pharmaceutical or cosmetic materials can be controlled by the adoption of one of two types of standard, or both standards. The first is a limit on the total numbers of viable organisms in a given weight or volume of material (total viable count, TVC), and the second is the exclusion of specific pathogens.

TVCs for different classes of organism, e.g. aerobic bacteria, yeasts/moulds or Enterobacteriaceae are described in both the EP (1997) and USP (1995), and although the microbial colonies which arise on the culture plates may be both numerous and varied in appearance, the counting procedures themselves and the interpretation of results are relatively straightforward. Consequently, viable counts are not considered further in this chapter, but are discussed in detail in Chapter 4. A worked example calculation can be found in Chapter 14.

The second type of test, in which named species must be undetectable in a given weight or volume of sample, might reasonably be applied to a wide range of potential pathogens, but in practice only a small number of organisms are the subject of such pharmacopoeial tests. In addition to being potentially hazardous to health, the selected organisms possess other properties of importance, e.g. they may be common contaminants of raw materials or they may be indicative of the quality of the manufacturing processes. Tests for 'specified micro-organisms', as they are described in the EP, are applicable to *Escherichia coli, Staphylococcus aureus, Pseudomonas aeruginosa*, salmonellae and clostridia, and are generally much more difficult to perform and interpret than viable counts.

*E. coli* is an organism commonly encountered in the gastrointestinal tract of mammals, and is detectable in their faeces. Some strains of *E. coli* exist as harmless commensals (natural inhabitants) in the intestine and some are pathogens which produce enterotoxins responsible for diarrhoeal disease. It is essential, therefore, that this organism is excluded from pharmaceutical materials in order to avoid the risk of infection, but the test for the absence of *E. coli* is perhaps most important because the organism is indicative of faecal contamination. Materials of 'natural' origin are thus frequently submitted to a limit test for *E. coli* because of their vulnerability to contamination. Gelatin, which has its origins in the slaughterhouse, is one of the best examples of such a material, but others such as the thickening agents acacia, alginates, tragacanth and sterculia would be similarly tested, as would the enzymes pancreatin, pepsin and trypsin, and miscellaneous products such as starch and digitalis preparations.

Salmonellae are of importance for much the same reasons as *E. coli* since they, too, commonly inhabit the intestine of animals. Their pathogenic potential is, however, substantially greater than that of *E. coli* since harmless strains of salmonellae are less frequently encountered and some strains can initiate infections from the ingestion of very small numbers of cells. Salmonellae, therefore, are the subject of microbial limit tests because they represent a major infection hazard rather than for any other reason, and this is reflected in the fact that they should not be detectable in larger sample sizes (typically 10 or 50 g) than those used for *E. coli* limit tests.

*Staphylococcus aureus* is similar to *E. coli* in that a wide variation in virulence is exhibited by different strains. Just as *E. coli* can inhabit the human intestine without doing any recognizable harm, so *Staphylococcus aureus* inhabits the human skin and nose without any apparent ill-effects. However, some strains of the species exhibit a marked pathogenic potential and can cause serious infections that may originate on the skin but progress to other anatomical sites, possibly with life-threatening consequences. There is another parallel with *E. coli* in that the appearance of the organism as a contaminant may be an indicator of poor product quality, because one of the most likely sources of *Staphylococcus aureus* in a manufactured product is the skin of the production personnel. Limit tests for *Staphylococcus aureus* are most likely to be applied to topical products, and the absence of the organism from such products is a USP requirement.

*Pseudomonas aeruginosa* has gained a great deal of attention for several reasons. It is a potential pathogen capable of causing infection at vulnerable sites in healthy persons, e.g. the eye, but it is particularly important as an opportunist pathogen which may cause infection in many regions of the body in persons with impaired immunity or underlying disease. Quite apart from this pathogenic potential, the organism would be a potential problem as a contaminant of pharmaceutical materials anyway because isolates are frequently found to be resistant to commonly used preservatives. Moreover, the organism has a low nutritional demand, i.e. it can use a wide variety of organic materials at low concentrations as nutrients and achieve relatively high cell densities. Water is one of the natural habitats of *Pseudomonas aeruginosa*, and stored water is more likely to be contaminated than that which is freshly purified. Oral antacid mixtures, e.g. aluminium hydroxide, have also been found to be susceptible to growth of *Pseudomonas aeruginosa*, and such products are frequently subjected to limit tests for this organism. Again, the USP requires that topical products are shown to be free of contamination with this species.

Clostridia have been included in the EP as the subject of an identification test since 1992, but there are no corresponding tests in the current USP. These organisms are anaerobes of which some species are pathogenic and others are considered to be harmless to humans. The EP (1997) describes a detection test to be applied to products where exclusion of pathogenic clostridia is essential. Talc and bentonite would be examples of such a material, although the test is not specifically included in either the BP (1998) or the EP (1997) monographs for these substances. There is also a semi-quantitative test for *Clostridium perfringens* which may be applied to products where the level of this species is a criterion of quality.

The situations in which limit tests for these five organisms are invoked varies between the different pharmacopoeias. The tests may be part of the monographs for particular materials, as in the BP and EP, or they may be invoked for certain classes of product, e.g. topicals, as in the USP. Tests for specified micro-organisms exhibit points of similarity and of contrast with tests for preservative efficacy. They are similar in that they act as a guide to manufacturers who wish, or are required by regulatory authorities, to include such tests in the specifications for the raw material or finished product. In contrast, however, tests for the absence of specified organisms are not intended solely or even primarily for application during product development, but they are to be used as an integral part of the quality assurance programme during manufacture.

## 6.3   General principles of the conduct of tests for specified micro-organisms

The aim of these tests is to produce evidence that the designated organisms are not detectable in a stated weight or volume of the material to be sampled. Just as it is impossible to design a test for sterility which can be *guaranteed* to detect any microbial contamination, it is also impossible to design a testing scheme which will, *without fail*, reveal the presence of specific bacteria. The ease with which the cells of a particular species might be detected will be influenced not only by the absolute concentration at which they are present in the sample but also by their concentration relative to other organisms. Thus, it may be easy to detect salmonellae when they represent half of a bacterial population which is very large anyway, but extremely difficult to do so if there are fewer than ten *Salmonella* cells among ten million non-salmonellae. The tests are therefore designed to maximize the chances of detecting the organisms in question, even when they are present in very low concentrations and numerically overwhelmed by other unrelated organisms.

There is another parallel with tests for sterility in that it is possible for the operators themselves inadvertently to introduce the organism in question during the conduct of the test. In order to minimize such contamination it is clearly desirable that detection tests are performed by personnel who possess good aseptic technique and have appropriate laboratory facilities available. The preambles to the pharmacopoeial tests refer to the avoidance of conditions which may adversely affect any micro-organisms revealed in the tests. Thus, the use of ultra-violet light to minimize atmospheric contamination should be avoided if there is a risk that it may destroy the organism being sought.

The tests generally have the following features in common (Tables 6.1 and 6.2):

1   The preparation of the sample such that the micro-organisms are uniformly dispersed in an aqueous medium.

2   Transfer of an appropriate volume of the aqueous dispersion to a liquid culture medium which acts as an enrichment culture. 'Enrichment' in this context means that the concentration of the 'target' organism is increased in absolute terms and usually also as a percentage of the total microbial population.

3   The liquid enrichment culture is used to inoculate agar media which are selective for the organism in question.

4   Any isolates which display the characteristic appearance or reaction of the target organism on the selective agar are then subjected to confirmatory biochemical or immunological tests which are highly, or absolutely, specific.

### 6.3.1  *Sample preparation*

The intention here is to create a uniform dispersion of the contaminating micro-organisms in a solution, suspension or emulsion of the material under test. This procedure should hydrate the organisms and bring them to a metabolically active state, as well as neutralize any preservatives or antimicrobial activity inherent in the sample.

Many pharmaceutical and cosmetic products are preserved simply by virtue of their anhydrous nature, since microbial growth cannot proceed without water. Drying does not necessarily kill micro-organisms, and their metabolic activity is merely suspended while they are in the dry state. Consequently, the organisms in anhydrous products in particular may need incubation at a temperature of 30–35°C for several hours before they are returned to a metabolically active state. It is important that this revival period is restricted to 2–5 h, otherwise there is the risk of over-growth by other contaminants, thus making detection more difficult.

**Table 6.1** Procedures recommended by the *European Pharmacopoeia* (1997) in tests for specified micro-organisms[†]

| Medium | Organism | | | | |
|---|---|---|---|---|---|
| | Clostridia | Escherichia coli | Salmonella | Pseudomonas aeruginosa | Staphylococcus aureus |
| Liquid enrichment | Reinforced medium for clostridia | MacConkey broth | Lactose broth followed by tetrathionate bile brilliant green broth | Casein soya bean digest broth | Casein soya bean digest broth |
| Agar media (primary test) | Columbia agar | MacConkey agar | Deoxycholate citrate agar, xylose lysine deoxycholate (XLD) agar and brilliant green agar | Cetrimide agar | Baird–Parker agar |
| Result(s) of secondary tests which confirm the presence of organism in question* | A negative catalase reaction by anaerobic Gram-positive bacilli | Production of indole at 44°C | Reactions characteristic of *Salmonella* on triple sugar iron agar | Positive oxidase reaction | Positive coagulase or deoxyribonuclease tests |

*Additional appropriate biochemical or serological tests are recommended for *E. coli*, salmonellae, *Pseudomonas aeruginosa* and *Staphylococcus aureus*.
[†]Note added in proof: The liquid enrichment steps and the secondary test for *Pseudomonas aeruginosa* described in the EP (2000) differ slightly from those tabulated.

**Table 6.2** Procedures recommended by the *United States Pharmacopoeia* (1995) in tests for specified micro-organisms

| Medium | Organism | | | | |
|---|---|---|---|---|---|
| | Currently there is no clostridia test | *Escherichia coli* | Salmonella | *Pseudomonas aeruginosa* | *Staphylococcus aureus* |
| Liquid enrichment | | Lactose broth | Lactose broth followed by fluid selenite–cystine medium and fluid tetrathionate medium | Fluid soybean casein digest medium | Fluid soybean casein digest medium |
| Agar media | | MacConkey agar followed by Levine eosin–methylene blue agar | Brilliant green agar, xylose lysine deoxycholate (XLD) agar and bismuth sulphite agar | Cetrimide agar | Vogel–Johnson agar or Baird–Parker agar or mannitol-salt agar |
| Result(s) of secondary tests which confirm the presence of the organism in question* | | | Reactions characteristic of *Salmonella* on triple sugar iron agar | Characteristic appearance on media for pigment detection and positive oxidase test | Positive coagulase test |

*Additional appropriate biochemical or serological tests are recommended for *E. coli*, salmonellae, *Pseudomonas aeruginosa*.

An essential feature of sample preparation is elimination of antimicrobial activity. This may be in the form of a preservative which is part of the formulation, or it may simply be removal of adverse conditions such as extremes of pH or osmolarity which act to protect the product from spoilage. It may be tempting to think that any product that has passed a preservative efficacy test during its development is unlikely to be contaminated with pharmacopoeial test organisms. However, strains used for preservative testing are often more preservative-sensitive than contaminants which arise during manufacture and product use, and it is quite feasible that such contaminants will merely be suppressed, but not killed, by the preservative. Elimination of antimicrobial activity is usually achieved by dilution of the sample so that the preservative is no longer at an effective concentration, or by the addition of specific neutralizing materials such as lecithin and polysorbate (Tween) during sample preparation. This aspect is described in detail in Chapter 2.

### 6.3.2  *Liquid enrichment media*

The prepared sample is inoculated into an appropriate liquid medium which will support the growth of the organism in question. If the enrichment medium is also selective, it may achieve an increase in the ratio of that organism to the rest of the microbial population. Casein soya bean digest broth medium (tryptone soya broth) which is recommended in the pharmacopoeias for the detection of *Pseudomonas aeruginosa* and *Staphylococcus aureus* is non-selective and will permit organisms other than these to grow. In contrast, MacConkey's medium, brilliant green bile broth, fluid selenite cystine medium and fluid tetrathionate medium all exhibit some degree of selectivity and may permit growth of *E. coli*, salmonellae and other organisms of intestinal origin while restricting the growth of non-intestinal organisms.

### 6.3.3  *Selective agar media*

A range of 'solid' media are employed for the limit tests (Tables 6.1 and 6.2), and these also exhibit varying degrees of selectivity for the organisms in question. In some cases the medium will support the growth of that organism and support few others, e.g. the selectivity of cetrimide agar for *Pseudomonas aeruginosa*; in other cases it will support a relatively wide range of organisms, e.g. MacConkey's agar will allow different intestinal bacteria to grow not just *E. coli* and salmonellae.

Frequently, growth is accompanied by the production of characteristic metabolic products, e.g. enzymes, acids or bases which may cause changes in the appearance of the medium adjacent to the bacterial growth, e.g. precipitation, clearing of opacity or change in colour of a pH indicator. The appearance of the cultures often changes progressively and sometimes relatively quickly during incubation, so it is important to examine the culture plates more than once during the recommended incubation period. In several cases acids are produced as a result of the early bacterial growth, and these acids are later neutralized by further metabolism, so the indicator in the agar may change entirely from acid to alkaline colour within as little as 5 h.

Correct interpretation of microbial limit tests will therefore be promoted by the microbiologist's awareness of the reasons for the changes in appearance. In some cases it would be misleading to examine the cultures before the recommended time, even though there may be vigorous growth in evidence; likewise, examination beyond the recommended time might also be misleading.

The EP and the USP differ with regard to the precision of the recommended incubation conditions; the EP is generally more precise and recommends 35–37°C for 18–24 h in the majority of cases, whereas the USP almost invariably uses the term 'incubate' which it defines as 30–35°C for a period of 24–48 h. It is essential that persons conducting and interpreting the tests recognize that the appearance of many of the cultures could vary dramatically between the extremes of these permitted temperature/time combinations.

Correct interpretation of the tests is facilitated also by appreciation of the following:

1   The susceptibility of the different media to deterioration and loss of selective or diagnostic properties due to incorrect preparation, overheating during sterilization or oxidation during storage prior to use.

2   The differences in formulations of the common media and the extent and significance of variations on pharmacopoeial recommended formulae. There are, for example, four different formulae for MacConkey's agar in the Oxoid range, and subtle but sometimes significant variations exist between ostensibly identical products of different manufacturers. It is worth noting that the pharmacopoeias permit the use of other formulations of the recommended media or possibly other media entirely, provided that they can be shown to have similar selective and nutritional properties.

3   The value of having available plates of *un*inoculated culture media which have been incubated for the same period of time as the inoculated plates. This helps in the assessment of changes in appearance which are due to: (i) incubation; and (ii) the organism itself.

4   The importance of regular use of authentic control cultures both to validate the procedures by confirming that they will detect low levels of the subject organism, and to have reference specimens available with which to compare suspect isolates. The EP procedures are more rigorous than those of the USP in this respect; the EP specifies 100 organisms as the number which should be detectable in order to validate the tests, whereas the USP protocol is imprecisely specified as 1 ml of a 1000-fold dilution of a 24-h broth culture (at least $10^5$ cells). (See Chapter 2 for a more detailed discussion of this aspect.)

### 6.3.4  *Confirmatory biochemical or immunological tests*

These tests usually have a very high, though not necessarily absolute, specificity for the organism in question by detecting the presence of a particular metabolic product, enzyme or antigen of the organism. The tests, which may be commercially available as a kit, usually require bacterial growth to be taken directly from a suspect colony on a selective agar plate or from a subculture of such a colony if the components of the selective agar cause interference. Again, known positive and negative control cultures should be included (see Chapter 5).

## 6.4   The detection of *Staphylococcus aureus*

### 6.4.1   *Baird–Parker medium*

This medium contains potassium tellurite and lithium chloride which, together, restrict the growth of most organisms other than *Staphylococcus aureus*. The selectivity is further enhanced by the inclusion of glycine and pyruvic acid, both of which tend to promote staphylococcal growth. The medium also contains egg yolk emulsion, which makes it opaque and slightly yellow in colour. The potassium tellurite is reduced by staphylococci to metallic tellurium, so the colonies after 24 h at 37°C are black, and in the surrounding agar there is a zone of clearing about 5 mm wide due to the action of proteolytic enzymes on the egg yolk. On further incubation phospholipase enzymes produced by *Staphylococcus aureus* liberate free fatty acids from the phospholipids in the egg yolk, and these precipitate in the agar to give a zone of opacity immediately surrounding the colony; this, in turn, is surrounded by the original clear proteolysis zone (Plate 1).

Most micrococci and coagulase-negative staphylococci grow more slowly than *Staphylococcus aureus* so that they display small, black colonies after 24 h and these have little or no clearing. The exception is *Staphylococcus saprophyticus* which may mimic *Staphylococcus aureus* with respect to both the proteolysis and the phospholipase. Some laboratory manuals suggest that the two organisms may be distinguished because the phospholipase reaction is slower with *Staphylococcus aureus*, but this is not a particularly reliable guide and any black colonies which exhibit opacity within a wider clear zone should be tested for coagulase. *Proteus* and *Bacillus* species may also grow on this medium but the growth is delayed, often brown rather than black in colour, and without proteolytic clearing during the first 24 h of incubation (Plate 2).

### 6.4.2   *Mannitol salt agar*

As staphylococci normally grow on the skin, they are uninhibited by the relatively high osmotic pressures found in sweat. This resistance may be exploited in the detection and identification of staphylococci by the incorporation of 7.5% sodium chloride into a general purpose culture medium, making it selective for this group and inhibitory to the growth of many other contaminants of pharmaceutical, cosmetic and food materials.

Mannitol is also included in the medium because there is a correlation between mannitol fermentation and the positive coagulase reaction which is definitive for *Staphylococcus aureus*. Most coagulase-positive staphylococci are mannitol fermenters, and will produce acid fermentation products which impart a yellow colour to the colony and the surrounding agar due to the presence of phenol red as a pH indicator (Plate 3). Coagulase-negative staphylococci grow on the medium, but these normally produce a pink coloration in the colonies and the agar (Plate 4).

### 6.4.3   *Vogel–Johnson agar*

This medium contains some of the features of mannitol salt and Baird–Parker media in that it contains mannitol and phenol red which are characteristic of the former, and the potassium tellurite, lithium chloride and glycine (but not the egg yolk emulsion) of the latter. The medium has more tellurite than Baird–Parker medium, and it is highly selective for staphylococci with few other organisms able to grow within a 24-h incubation period. *Staphylococcus aureus* colonies are black and may exhibit a yellow halo in the surrounding agar due to the fermentation of mannitol, although this is not invariably present at 24 h (Plate 5). Colonies of coagulase-negative staphylococci are black and the surrounding medium is normally unchanged in colour, although some coagulase-negative species may exhibit mannitol fermentation on prolonged incubation.

### 6.4.4  *Coagulase test*

Coagulase is the name given to a variety of bacterial enzymes that promote clotting of human or rabbit plasma. Such enzymes may be produced by bacteria other than staphylococci, but staphylocoagulases are by far the most commonly studied and used for diagnostic purposes. *Staphylococcus aureus* is certainly the most frequently encountered coagulase-positive *Staphylococcus*, although there are one or two other species (rarely encountered animal isolates) which produce coagulase.

The coagulase enzyme may be either free, i.e. diffusible into the surrounding medium, or bound to the cell wall in which case it is often referred to as 'clumping factor'. The two types are detected by different tests; free coagulase promotes the change from fibrinogen to fibrin and forms a gel or clot when a saline suspension of bacterial cells is mixed with plasma in a test tube at 37°C, whereas bound coagulase is usually detected by agglutination of cells on a microscope slide. This slide test is simpler to perform and gives a rapid result because agglutination is usually apparent within a matter of 30 s. The tube test may require incubation for several hours before the clot is apparent. The two tests do not correlate completely.

There are several commercially available test kits based upon the slide coagulase test which employ latex beads coated with plasma. Fibrinogen bound to the latex detects the coagulase because a suspension of the beads rapidly agglutinates when mixed with bacterial cells from a *Staphylococcus aureus* colony (Plate 6). These kits are both easy to use and reliable and have, in most laboratories, largely replaced the traditional slide and tube methods.

### 6.4.5  *Deoxyribonuclease test*

This is an alternative to the coagulase test which is recommended in the EP for the detection of *Staphylococcus aureus*. There is a strong correlation between the production of DNAse and the coagulase reaction among staphylococci, although DNAse is found in more *Staphylococcus* species than coagulase. Coagulase is also possessed by other organisms, including *Streptococcus pyogenes* and *Serratia marcescens*. DNAse is detected by inoculating the suspect organism onto the surface of an agar medium containing tryptose and DNA. Following incubation for 24 h at 37°C, the plate is flooded with 0.1 M hydrochloric acid; this precipitates the intact DNA to cause opacity in the agar. If the DNA has been digested, there is a clear zone of several millimetres width surrounding the bacterial growth (Plate 7).

## 6.5   The detection of *Pseudomonas aeruginosa*

### 6.5.1   *Cetrimide agar*

Cetrimide is a quaternary ammonium antiseptic which has a broad spectrum of activity against many bacterial contaminants of pharmaceutical and cosmetic materials. The major gap in its antimicrobial spectrum is *Pseudomonas* species, so inclusion of cetrimide in a medium will render it selective for these organisms. The selectivity is not absolute, however, and although the medium is recommended in the pharmacopoeias for the detection of *Pseudomonas aeruginosa* it will also support the growth of other *Pseudomonas* species.

Usually, growth arises as small colonies which may be colourless when they are first visible, but the pigments that are characteristic of *Pseudomonas* species usually start to develop by 24 h, and their intensity increases with further incubation (Plate 8). Occasionally non-pigmented strains are encountered, so the absence of pigment cannot be taken to indicate that the organism is not *Pseudomonas*; indeed, environmental strains of *Pseudomonas aeruginosa* are frequently non-pigmented. A small number of other bacteria are capable of growth on cetrimide agar including *Aeromonas*, *Alkaligenes* and *Proteus* species (Plate 9).

### 6.5.2   *Pseudomonas* media for the detection of fluorescein and pyocyanin

Since over 90% of *Pseudomonas aeruginosa* isolates produce pigments (Holt *et al.*, 1994), this characteristic is of value in identification of the species, and the USP recommends the use of two media for the purpose. The two pigments most commonly encountered are fluorescein, which is yellow and fluoresces under ultra-violet light, and pyocyanin which is blue/green and unique to this particular species. Some strains of *Pseudomonas aeruginosa* also produce pigments which are red or brown (Plate 10), but these are much less common.

The cetrimide agar formulae in the USP and the EP are identical, and the USP recommends cetrimide agar base for the detection of pyocyanin. For fluorescein production the medium differs in its protein basis and in the substitution of 0.15% potassium dihydrogen phosphate for the 1.0% potassium sulphate. Several species of *Pseudomonas* produce fluorescein or other fluorescent pigments, so fluorescence under ultra-violet light does not confirm that the species is *Pseudomonas aeruginosa*.

### 6.5.3   *Oxidase test for* Pseudomonas aeruginosa

*Pseudomonas aeruginosa* is incapable of fermentation, and its metabolism is respiratory in nature. It therefore possesses cytochrome oxidase for the purpose of transferring electrons directly to molecular oxygen in order to form water at the end of the metabolic pathway. This oxidase can be detected by exposing the bacterial cells to a dye which is an artificial electron donor and changes colour as it is oxidized on transfer of electrons to the oxidase. Such a dye is tetramethyl-*p*-phenylenediamine dihydrochloride which, as a 1.0% aqueous solution, is known as Kovac's oxidase reagent.

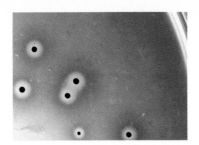

Plate 1  *Staphylococcus aureus* on Baird–Parker medium (48 h at 37°C), showing typical black colonies surrounded by zones of opacity due to fatty acid precipitation, beyond which are clear zones of proteolysis.

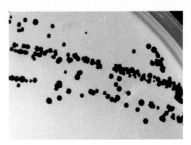

Plate 2  *Proteus mirabilis* on Baird–Parker medium (48 h at 37°C), showing black colonies but no precipitation or proteolysis. The appearance does not change significantly on further incubation.

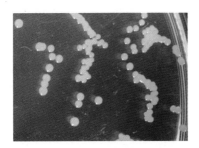

Plate 3  *Staphylococcus aureus* on mannitol salt agar (24 h at 37°C), showing yellow coloration in the phenol red indicator due to acid fermentation products of mannitol.

Plate 4  *Staphylococcus epidermidis* (a coagulase-negative *Staphylococcus*) on mannitol salt agar (24 h at 37°C). Like many coagulase-negative species, it does not normally ferment mannitol.

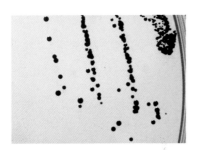

Plate 5  *Staphylococcus aureus* on Vogel–Johnson medium (24 h at 37°C), showing typical black colonies. Some reference sources indicate that a distinct yellow zone due to acid fermentation products of mannitol should also be seen, but this is not invariably present.

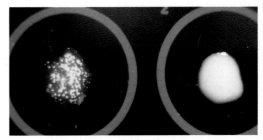

Plate 6  *Staphylococcus aureus* (coagulase-positive, left) showing agglutination, and *Staphylococcus epidermidis* (coagulase-negative, right) exhibiting no agglutination in a commercially available kit form of the coagulase test.

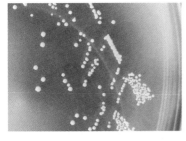

Plate 7  *Staphylococcus aureus* exhibiting hydrolysis of deoxyribonucleic acid in DNA-containing agar, as shown by a clear zone surrounding growth after flooding the plate with hydrochloric acid which precipitates unhydrolysed DNA (24 h at 37°C).

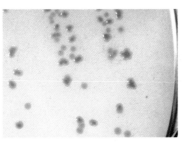

Plate 8  *Pseudomonas aeruginosa* on cetrimide agar (24 h at 37°C), showing typical pigmentation.

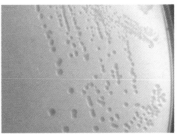

Plate 9  *Proteus vulgaris* on cetrimide agar (24 h at 37°C). This and other species of *Proteus* may grow, but produce no pigment.

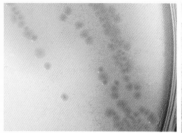

Plate 10 *Pseudomonas aeruginosa* exhibiting the somewhat less commonly encountered orange-brown pigmentation on cetrimide agar (24 h at 37°C).

Plate 11 *Pseudomonas aeruginosa* on cetrimide agar immediately (left), 15 s (centre) and 30 s (right) after addition of oxidase reagent.

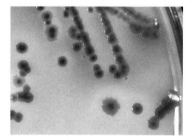

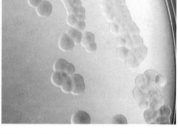

Plate 12 *Escherichia coli* on MacConkey's agar (24 h at 37°C), showing the typical pink colonies due to acid accumulating from fermentation of lactose in the medium and changing the colour of the neutral red indicator. Precipitation of bile in the agar due to the low pH is also evident.

Plate 13 *Salmonella arizonae* on MacConkey's agar (24 h at 37°C), showing the typical appearance of non-lactose fermenters.

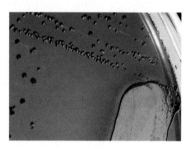

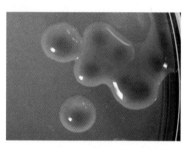

Plate 15 Large mucoid colonies of *Klebsiella pneumoniae* on eosin–methylene blue agar (24 h at 37°C). Acid production is not sufficiently vigorous to precipitate the stain.

Plate 14 *Escherichia coli* on eosin–methylene blue agar (24 h at 37°C), showing the typical green metallic sheen due to acid precipitation of the eosin–methylene blue complex.

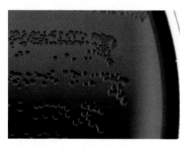

Plate 16 *Salmonella arizonae* on eosin–methylene blue agar (24 h at 37°C), showing colourless translucent colonies typical of non-lactose fermenters.

Plate 17 The detection of indole in tryptone water cultures (24 h at 44°C) of *Escherichia coli* (indole-positive, left) and *Klebsiella pneumoniae* (indole-negative, right).

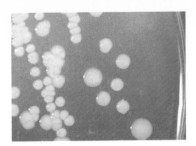

Plate 18  *Enterobacter aerogenes* on brilliant green agar (24 h at 37°C), showing yellow coloration in the phenol red indicator due to lactose fermentation.

Plate 19  *Salmonella arizonae* on brilliant green agar (24 h at 37°C). The colour of the medium is typical of that resulting from growth of a non-lactose fermenter.

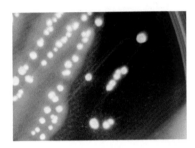

Plate 20  *Edwardsiella tarda* on XLD agar (24 h at 37°C), showing the alkaline reaction in the agar and the blackening of the colonies which is normally exhibited by species of *Edwardsiella*, *Salmonella* and, less commonly, *Citrobacter*.

Plate 21  *Citrobacter freundii* on XLD agar (24 h at 37°C photographed against a dark background), showing acid production and precipitation in the medium typical of organisms capable of fermenting the constituent sugars.

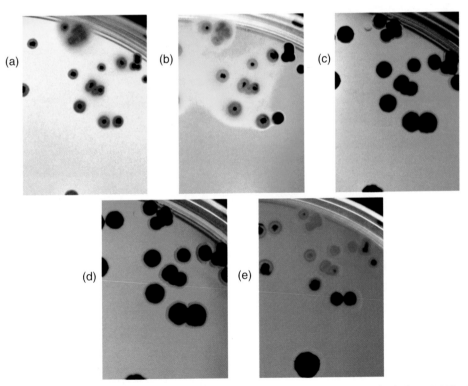

Plate 22  A strain of *Citrobacter freundii* on XLD agar incubated at 37°C and photographed after (a) 21 h, (b) 27.5 h, (c) 40 h, (d) 49.5 h and (e) 72 h. In this strain the acid which resulted in the yellow coloration at approximately 24 h was subsequently converted to an alkaline product with an associated gradual loss of the hydrogen sulphide-induced black coloration of the colonies.

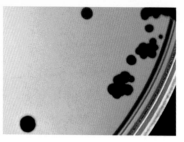

Plate 23   *Salmonella abony* on XLD agar (24 h at 37°C). Note the similarity in appearance to *Citrobacter freundii* (Plate 22d).

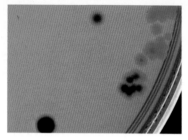

Plate 24   The same culture of *Salmonella abony* as that in Plate 23, but photographed after a further 24-h incubation. Note that it is possible for the black precipitate to disappear almost completely.

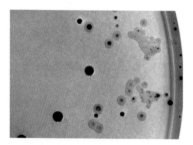

Plate 25   *Salmonella abony* on bismuth sulphite agar (24 h at 37°C). The colonies may be black or green, even within a pure culture.

Plate 26   *Klebsiella pneumoniae* on bismuth sulphite agar (72 h at 37°C). Normally non-salmonellae are inhibited, but heavy inocula may promote growth which may then even display the metallic sheen which is frequently described as characteristic of salmonellae.

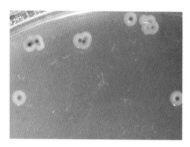

Plate 27   *Salmonella abony* on deoxycholate citrate agar (24 h at 37°C). The medium shows the alkaline yellow-orange colour of the neutral red indicator, and the colony has a central grey dot due to hydrogen sulphide production.

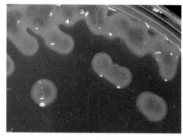

Plate 28   Mucoid colonies of *Klebsiella pneumoniae* on deoxycholate citrate agar (24 h at 37°C) showing acid production indicated by the pink coloration.

Plate 29   Triple sugar iron agar showing four different reactions plus an uninoculated tube for comparison (24 h at 37°C). From the left, *Escherichia coli* with acid in slant and butt, *Proteus vulgaris* with acid slant and butt plus $H_2S$ production, uninoculated control, *Pseudomonas aeruginosa* with alkaline slope and butt, *Salmonella arizonae* with acid butt, alkaline slope and $H_2S$ production.

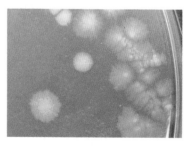

Plate 30   *Clostridium perfringens* on Colombia agar (24 h at 37°C).

The test is performed by adding the reagent dropwise to young isolated colonies growing upon the agar surface, or by adding a loopful of material from the colony onto a filter paper stick soaked in reagent. A positive reaction is indicated by the reagent becoming blue/purple very quickly. Usually the colour change is apparent after 10 s, but it may sometimes take up to 30 s (Plate 11). The development of a blue colour after a minute or more is of no significance because the colonies of oxidase-negative organisms will gradually become blue, as, indeed will the agar itself, due to atmospheric oxidation of the reagent (which should therefore be prepared immediately before use). If the reagent is added to old cultures which are deeply pigmented the colour development in the reagent is more difficult to see against the background colour.

*Pseudomonas aeruginosa* is by far the most common oxidase-positive contaminant of pharmaceutical and cosmetic materials, and one of the few which will also grow on cetrimide agar. *Bacillus* and *Aeromonas* species may arise as pharmaceutical contaminants and isolates of these may also exhibit a positive oxidase reaction, although if it is seen at all in *Bacillus* isolates it is usually weak. The oxidase test is particularly useful for distinguishing *Pseudomonas aeruginosa* from members of the Enterobacteriaceae because the latter are all oxidase-negative.

## 6.6   The detection of *E. coli* and Enterobacteriaceae

### 6.6.1   *Introduction*

The Enterobacteriaceae is a family of Gram-negative, rod-shaped bacteria which includes *E. coli* – the only important species within the genus *Escherichia* – together with species of *Klebsiella, Enterobacter, Salmonella, Citrobacter, Edwardsiella, Shigella* and *Proteus*, among others. They have in common the characteristics that they can metabolize sugars and other carbohydrates either by a respiratory (oxidative) or a fermentative metabolism. Since fermentation is, strictly speaking, an anaerobic process this means that all Enterobacteriaceae are facultative anaerobes, i.e. are capable of growing in both aerobic and anaerobic conditions. Unfortunately, the term 'fermentation' is often used rather loosely in microbiology simply to describe the metabolism of sugars or other carbohydrates to give organic acids which are detected by a pH indicator in the medium. *Pseudomonas* species are incapable of fermentation, thus they are not members of the Enterobacteriaceae and are described as 'non-fermenters'.

The genera which constitute the Enterobacteriaceae exhibit wide variations in their ability to ferment different sugars, but the ability to ferment lactose is of particular importance as a means of distinguishing between them. *E. coli* and species of *Klebsiella* and *Enterobacter* ferment lactose, whereas *Salmonella, Shigella* and *Proteus* species do not. The term coliforms is often used to describe the lactose-fermenters, although unfortunately this term also has sometimes been used loosely to describe any Gram-negative enteric bacterium.

*E. coli* is important because it is an inhabitant of the mammalian intestine, and its presence in a sample may be indicative of faecal contamination. It is desirable therefore to distinguish *E. coli* from the other coliforms such as *Klebsiella pneumoniae*: this was previously called *Aerobacter aerogenes* and, more recently, *Klebsiella aerogenes*. Distinction between this species, other coliforms and *E. coli* is most conveniently achieved by a test for the metabolic product indole, which only *E. coli* can produce when incubated at 44°C.

### 6.6.2 *MacConkey agar*

Several formulations of this medium are available with differing degrees of selectivity. The version described in the EP and USP is the most selective, and it will permit the growth of coliform bacteria and Gram-negative, non-lactose-fermenting gut bacteria while excluding all Gram-positive cocci; the latter may appear on the less-selective formulae. The basis of selectivity is the inclusion of both bile salts which inhibit organisms not capable of growing in the intestines, and of crystal violet to inhibit cocci. Thus all the Gram-negative enteric pathogens (*Salmonella* and *Shigella* for example) and the gut commensals may grow on MacConkey's agar, as well as organisms such as *Pseudomonas aeruginosa* which are less commonly found in the gastrointestinal tract.

The length of Lactose is included as the fermentable carbohydrate, and in agar formulations of MacConkey's medium the acid products of fermentation are detected using neutral red; bromocresol purple is used as an alternative indicator in broth media. Neutral red is pink in acid and orange/brown in alkaline conditions, so *E. coli*, *Klebsiella*, *Enterobacter* and other lactose-fermenters produce red colonies and impart a red colour to the surrounding medium after 24 h at 37°C. The low pH may result in precipitation of the bile, so opacity in the agar surrounding the growth may occur (Plate 12). *Klebsiella* species may produce mucoid colonies which appear sticky and shiny due to production of a polysaccharide capsule or slime layer, and this helps to distinguish them from *E. coli*; however, non-mucoid strains also exist so this feature is not an infallible guide to identity. Mucoid colonies of *Klebsiella* species on MacConkey's agar are similar in appearance to those on deoxycholate citrate agar which utilizes the same indicator (see Plate 28). Non-lactose fermenters grow on the medium to give colourless colonies, and the agar itself may appear slightly more orange/brown than that in an uninoculated plate after 24 h at 37°C (Plate 13).

The length of incubation is important with MacConkey's agar because the acid products of lactose fermentation may be further metabolized to give a more alkaline reaction to the medium on prolonged incubation. Hence colonies that are distinctly red after one day may be less so, or even colourless, after two or more days. If no red colonies have appeared after 48 h incubation at 37°C they are extremely unlikely to do so on further incubation.

### 6.6.3 *Crystal violet, neutral red bile agar with glucose and lactose*

This medium is recommended by the EP for the detection of the Enterobacteriaceae and it is, in most respects, similar to MacConkey's medium. It contains bile salts and crystal violet as the selective agents which suppress the growth of Gram-positive and non-enteric bacteria, lactose and a neutral red indicator. The non-lactose-fermenting organisms that would normally grow on MacConkey's agar to give colourless colonies have, on this medium, glucose available as an alternative substrate for acid production, so that all colonies of Enterobacteriaceae which arise will appear pink.

## 6.6.4 *Levine eosin–methylene blue agar*

This medium is intended primarily to distinguish *E. coli* from other Gram-negative enteric bacteria on the basis of their abilities to ferment lactose, and is recommended by the USP but not the EP. The medium as originally described by Levine contained lactose as the only sugar, but subsequent modifications also contained sucrose, and these too are available commercially. Eosin and methylene blue act as inhibitors of both Gram-positive bacteria and some of the more sensitive Gram-negative species.

E. coli ferments lactose to pyruvic acid and imparts such a low pH that an eosin–methylene blue complex is precipitated in the agar; this results in *E. coli* colonies exhibiting a green metallic sheen when viewed by reflected light (Plate 14); a blue-black coloration is seen by transmitted light. Other Gram-negative lactose fermenters do not generally produce sufficient acid to bring about this precipitation, so there is no metallic sheen and the colonies appear purple, often with darker, even black, centres (Plate 15). Non-lactose fermenters usually produce colourless translucent colonies which show little contrast against the background of the medium itself (Plate 16).

## 6.6.5 *Indole test for confirmation of* E. coli

*E. coli* can modify the side chain of the amino acid tryptophan to produce indole. This ability can be detected by the use of a reagent which gives a characteristically coloured product on reaction with indole. Kovac's reagent, which is approximately 5% (w/w) *p*-dimethyl-aminobenzaldehyde in acidified amyl or *iso*-amyl alcohol is the most common, and Ehrlich's reagent uses the same chromophore; *p*-dimethyl-aminocinnamaldehyde is more sensitive, but paradoxically less frequently employed.

Indole production is also seen in some species of *Proteus* and is very occasionally encountered in other genera, but only *E. coli*, produces indole during growth at 44°C. The test is therefore normally performed at this temperature by inoculating a tube of tryptone water, a culture medium with a high tryptophan content, with a suspect colony and incubating for 24 h in a thermostatically controlled water bath; the temperature control on an incubator may not be sufficiently precise to give a reliable result. A few drops of reagent are added to the tryptone water culture which is gently shaken and a bright red colour (blue with *p*-dimethyl-aminocinnamaldehyde) appearing immediately or within a few seconds confirms the presence of *E. coli* (Plate 17). A negative result is indicated by shades of brown or orange. Kovac's reagent is more convenient to use than Ehrlich's because the solvent, amyl alcohol, is immiscible with the aqueous culture medium and forms the upper layer in which the colour is concentrated. Ehrlich's reagent is based upon ethanol, so it is necessary to add a few drops of xylene to the tube in order to achieve the same effect.

## 6.7   The detection of *Salmonella* species

## 6.7.1 *Introduction*

The detection of salmonellae can be more problematic, as several relatively harmless organisms with much the same metabolic profile as salmonellae can easily be mistaken for them if reliance is placed solely on the use of selective media. This difficulty is reflected both by the relatively large number of media that are commercially available for salmonellae detection, and by the fact that two or more are normally used simultaneously. The EP recommends the use of 'at least two' media out of the three described [brilliant green agar, xylose lysine deoxycholate (XLD) agar and deoxycholate citrate agar], whereas the USP directs that all three (brilliant green, XLD and bismuth sulphite agars) should be employed. Both pharmacopoeias require that suspect isolates from these media should then be inoculated into triple sugar iron agar to provide further evidence of a positive identification.

Despite the use of multiple selective solid media it is still possible for non-salmonellae to arise which conform in most, if not all, of the features which the pharmacopoeias describe as characteristic of salmonellae. *Edwardsiella* species for example are rarely encountered as human pathogens but are frequently indistinguishable from *Salmonella* species when grown on the media mentioned above, and both *Citrobacter* and *Proteus* species bear sufficient resemblance to cause problems. The problem of distinguishing between these organisms is compounded by the fact that the pharmacopoeial descriptions of salmonellae on the recommended media are of minimal value because of their brevity.

A positive identification of *Salmonella* is usually only recorded if the isolated organism not only conforms to the standard description on selective media but the identity is confirmed by a commercially available test kit (API or similar) and the culture shows positive agglutination with *Salmonella* antiserum.

### 6.7.2  Brilliant green agar

This medium contains the triphenylene diamine dye, brilliant green, as the selective agent. It is inhibitory to Gram-positive bacteria, many yeasts and moulds and, to a certain extent, some intestinal pathogens at the concentrations used in the medium. The growth of coliforms is suppressed although heavy inocula will certainly initiate some growth. It is primarily salmonellae which will grow readily although, even for these organisms, a degree of inhibition may be seen, particularly with low inocula. Despite the name the medium does not appear green because it contains phenol red, which represents the dominant colour.

Lactose and sucrose are present so growth of *E. coli*, *Klebsiella*, *Enterobacter* and other organisms capable of fermenting either of these sugars may result in a medium which is green or yellow depending upon the amount of acid produced (Plate 18). All non-fermenters that are not inhibited by the brilliant green (not just *Salmonella* species) will produce colonies which are red or pink (Plate 19). The colour also extends to the medium, so the whole plate may be similarly coloured, in which case the individual colonies may not appear to have a particular colour of their own because there is no contrast with the medium upon which they are growing. Non-sucrose-fermenting species of *Citrobacter* and *Proteus*, as well as species of *Edwardsiella* and *Pseudomonas*, may grow on brilliant green agar to give colonies that are indistinguishable from those of salmonellae.

### 6.7.3  Xylose, lysine, deoxycholate (XLD) agar

This is probably the most useful and discriminating of the four recommended salmonellae media, but it is also the most difficult to interpret. The medium was originally intended for the isolation of shigellae from faeces, but it has since been developed for the detection of both shigellae and salmonellae. It contains sodium deoxycholate at half the concentration used in deoxycholate citrate agar (see below), and thus it is less inhibitory for coliform bacteria; consequently most of the Enterobacteriaceae have the potential to grow on XLD medium.

XLD agar contains three sugars, xylose, lactose and sucrose, plus phenol red as indicator. If none of the sugars is fermented (as with *Shigella* and some *Proteus* species), the organism will produce translucent colonies and the medium colour will be unchanged or become slightly redder; Plate 20 shows the typical alkaline colour of the medium. If any one of the sugars is fermented, and there is no reversal of the lowered pH, then the colonies become, and stay, yellow (*E. coli, Klebsiella, Enterobacter, Citrobacter* and some *Proteus* species; Plate 21).

The amino acid lysine is included because organisms such as *Salmonella* and *Edwardsiella* can decarboxylate it, neutralize the acid from sugar fermentation, and change the medium colour from yellow to red; this usually occurs only after approximately 18–20 h incubation at 37°C. Medium colour, therefore, will not distinguish salmonellae from the non-fermenters like *Pseudomonas* since it will be red in both cases. *E. coli* and other lactose or sucrose fermenters may also decarboxylate lysine, but this causes no confusion because the sugar concentration so far exceeds that of the lysine that the acid fermentation products cannot be neutralized and the medium remains yellow throughout.

The medium also contains sodium thiosulphate which may be reduced by some organisms to give hydrogen sulphide; this reacts to give a black precipitate with the ferric ammonium citrate which is also present. Thus, hydrogen sulphide-producers appear as colonies which may be either yellow with black centres (e.g. *Citrobacter*; Plate 22) or red with black centres (*Salmonella*; Plate 23). Confusion can be caused by some strains of *Citrobacter* which initially produce acid products to give a yellow coloration, but then proceed to neutralize the acid within a 48-h incubation period. Plate 22 shows the same Petri dish photographed five times over 72 h. It is apparent that the medium colour and hydrogen sulphide production make the culture more closely resemble *Salmonella* as the time proceeds.

Both the EP and the USP recognize that *Salmonella* colonies do not always produce sufficient hydrogen sulphide for the blackening to be apparent, so the official description is 'red, with or without black centres'. It is important to appreciate that the blackening is a complex phenomenon which is time-, temperature- and pH-dependent. Hydrogen sulphide production is retarded in alkaline conditions, so on prolonged incubation (beyond 48 h at 37°C) the black colour may diminish or disappear (contrast Plates 23 and 24). It should also be remembered that the hydrogen sulphide precipitate may entirely obscure the yellow or pink *colony* colour, so the colony (as distinct from the surrounding medium) is entirely black.

### 6.7.4  *Bismuth sulphite agar*

This medium is currently recommended in the USP for the detection of salmonellae, and was recommended also by the BP until 1988 when it was replaced by XLD. It was originally developed for the detection of *Salmonella typhi*, which, of course, is now a very unlikely contaminant of pharmaceutical or cosmetic materials. This fact, together with the relative difficulty in preparation and short shelf-life of the medium, are factors which make alternatives appear preferable. This agar contains both bismuth sulphite and brilliant green which, together, inhibit the growth of Gram-positive bacteria and coliforms; thus, it has a high degree of selectivity, and few organisms other than *Salmonella* species will grow well from low inocula.

Even salmonellae however, are not totally resistant to the activity of bismuth salts, and these also may be inhibited to a degree which depends upon the age of the prepared medium. The inhibition is greatest in freshly poured plates (which appear white due to reduction of the brilliant green dye), and these should be used if the *Salmonella* concentration is expected to be very high. After 3–4 days in the refrigerator the green colour is restored by atmospheric oxidation and the proportion of cells giving colonies is much increased. The instructions given by the suppliers for preparation of the medium may suggest that thick plates should be poured (25 ml rather than 15–20 ml) and the medium should be used immediately, but these instructions are more pertinent to heavy inocula, e.g. faecal specimens. The influence of ageing of the medium on the degree of growth inhibition sometimes allows the growth of *Citrobacter* and other species to be observed on old plates.

The appearance of *Salmonella* species on bismuth sulphite agar may be quite variable, although the USP simply describes it as 'black or green' (Plate 25). The sulphite from the bismuth salt acts as a substrate for hydrogen sulphide production, so salmonellae – and indeed any organisms that can grow on the medium and possess this reducing capacity – may produce colonies which are black in colour because the medium also contains ferrous sulphate with which the hydrogen sulphide reacts. The black colour may appear as a central spot in the *Salmonella* colony (described in some textbooks as a rabbit's eye), or the whole colony may be black. Reduction of the sulphite leaves the bismuth as a precipitate; this may give a metallic sheen or mirror on the surface of the medium surrounding the colony, and this is usually more prominent when the colony in question is well isolated. Areas of confluent growth may show no sheen at all, nor even much blackening, and simply appear green. It should also be recognized that a metallic sheen is neither unique to salmonellae nor does it *always* arise in *Salmonella* cultures. Indeed, it may even be displayed occasionally, particularly on prolonged incubation, by some species which do not normally grow well on this medium, e.g. *Klebsiella pneumoniae* (Plate 26).

Organisms other than *Salmonella* sp. may occasionally give colonies having one or more of the above characteristics but few, if any, grow as vigorously. *Citrobacter* is capable of hydrogen sulphide production and may give colonies that are very similar in appearance to those of *Salmonella* but usually do not exhibit the metallic sheen. *Proteus* species may also give brown or green colonies, and coliforms, which are rarely encountered, are most commonly green.

### 6.7.5  *Deoxycholate citrate agar*

This medium, too, is available in several different formulations and the precise formula described in the EP (the medium is not recommended in the USP) is unavailable from the major British suppliers. In principle, the medium is similar to MacConkey's agar, in that both achieve selectivity by the incorporation of bile, both contain lactose and neutral red indicator, and both media are therefore selective for Gram-negative intestinal bacilli. Deoxycholate citrate agar (DCA) contains a specific bile salt, sodium deoxycholate, and the concentration employed determines the degree of selectivity achieved. The various modifications of the medium differ with respect to the concentrations of deoxycholate, sodium citrate, presence of thiosulphate and the nature of the ferric salt employed. The relatively high concentration recommended in the EP formula for DCA, together with the high concentration of sodium citrate, makes the medium more selective than MacConkey's agar, and so coliforms and *Proteus* species are less likely to appear on DCA.

Some formulations of DCA (but not that recommended in the EP) include a low concentration of sodium thiosulphate. This may be reduced by some enteric organisms (*Salmonella, Citrobacter, Proteus* for example) to give sulphides which in turn react with the ferric salt in the medium to impart a central grey or black coloration to the colonies of the organisms in question (Plate 27). In the absence of a reducible source of sulphur in the EP formula, colonies grown on this specific medium will either be colourless (*Salmonella, Shigella, Proteus, Pseudomonas*) or pink (*E. coli, Klebsiella, Enterobacter*) due to lactose fermentation (Plate 28). As with MacConkey's medium, vigorous lactose fermenters may reduce the pH in the agar sufficiently to cause precipitation of the bile salt.

### 6.7.6  *Triple sugar iron agar*

This medium is designed to aid the detection of Enterobacteriaceae, and because it contains no selective agents it will support the growth of a wide range of bacteria. The mere presence of growth on the medium, therefore, has little or no diagnostic significance; it is the characteristics of the growth which are important. It follows from this that the medium must be inoculated with a pure culture, often an isolated colony from another identification medium, and no useful information is gained if it is inoculated with a mixed culture.

There are several different reactions which can be detected on triple sugar iron agar, but it is important that the medium is correctly prepared in order to provide the maximum amount of diagnostic information. It should be poured as a slant (also called a slope) in a tube which is set at an angle such that the area of the surface (or slant portion) is increased so that its longest dimension is approximately the same as the depth of the lower portion (called the butt or deep) at approximately 2.5–3 cm; this results in the surface being aerobic and the butt relatively anaerobic. Tubes are preferable to screw-capped containers because the reactions which characterize growth of different organisms are dependent upon oxidative reactions, and oxygen access to the agar surface is better in an unsealed tube.

The medium is designed to distinguish organisms on the basis of their abilities to ferment the sugars, dextrose, lactose and sucrose, and this is indicated by the presence of phenol red. Gas production is another characteristic referred to in the manuals of culture media suppliers, and this means carbon dioxide gas from sugar fermentation rather than hydrogen sulphide gas from the reduction of thiosulphate, which is another medium component. The former is recognized simply as bubbles of gas between the agar and the wall of the tube or within the agar itself. The carbon dioxide production may be sufficient literally to split the agar into two or more sections. Hydrogen sulphide is detected as blackening of the agar due to its reaction with the ferrous sulphate in the medium, and this may be so extensive that any colour due to the phenol red is obscured.

Organisms which cannot ferment any of the sugars (e.g. *Pseudomonas aeruginosa*) oxidatively decarboxylate peptides to give amines at the agar surface, and these alkaline products diffuse throughout the medium which may, therefore, be uniformly pink (Plate 29). Such a result is sufficient to distinguish the organism from the Enterobacteriaceae.

Glucose is present at one-tenth of the concentration of both lactose and sucrose. If an organism can *only* ferment glucose, the organic acids produced confer a yellow colour throughout the medium during the early stages of incubation. However, when this sugar is exhausted the amines derived from peptide decarboxylation accumulate to cause reversion of the surface colour to pink. In the relatively anaerobic butt this reaction is minimal and the butt remains yellow, so a yellow butt and pink slant after 24 h is the appearance typical of organisms such as *Salmonella* and *Shigella* species (Plate 29). These two genera can then be distinguished on the basis of hydrogen sulphide production because most species of the former produce it, but none of the latter. As hydrogen sulphide production is inhibited in alkaline conditions it is normally assumed that tubes have an acid butt if they show such extensive blackening that the phenol red colour is obscured.

If the organism can ferment lactose *or* sucrose in addition to the glucose (*E. coli*, *Klebsiella*, *Enterobacter* and some *Proteus* species), the resultant acids accumulate in such high concentration that they cannot be neutralized and both the slant and the butt are yellow after 18–24 h (Plate 29). Such a reaction is sufficient to exclude *Salmonella* and *Shigella* species as the organism in question because neither of these ferment lactose or sucrose.

It should be apparent from the above description that the incubation period is critical for the correct interpretation of the results, and it should also be recognized that in this medium organisms that are not *Salmonella*, e.g. some *Proteus* species and *Edwardsiella*, can give reactions which conform to it in every respect.

### 6.7.7  *Confirmation of salmonellae*

The fact that the EP recommends the use of three different selective media followed by triple sugar iron agar and 'appropriate biochemical and serological tests' to detect the presence of salmonellae indicates how difficult such organisms are to detect with confidence. The USP is somewhat less demanding in that the wording leaves further testing after triple sugar iron agar to the discretion of the operator. 'Appropriate biochemical tests' is usually taken to mean a presumptive identification of the organism using a biochemical testing kit such as API 20E system. Alternatively or additionally, the organism can be investigated for its conformation to the standard pattern of results expected of salmonellae when subjected to a panel of tests which are particularly discriminating for these organisms; such tests are described in the standard textbooks and in the manuals of the culture medium suppliers, e.g. the Difco Manual.

The serological (immunological) identification of salmonellae is an extremely complicated subject, with a bewildering array of antigen and antisera preparations available commercially for the purpose, and is best conducted by a specialist laboratory.

## 6.8  The detection of clostridia

The EP describes a test for the absence of clostridia in materials where exclusion of pathogenic *Clostridium* species is essential. The prepared sample is divided into two identical parts, one of which is heated at 80°C for 10 min to kill non-spore-forming organisms, while the other is left unheated. The liquid enrichment culture stage is achieved using reinforced clostridial medium; this is a broth which is rendered semi-solid by the inclusion of $0.5 \, gl^{-1}$ agar in order to reduce its viscosity and impede oxygen diffusion, and has added cysteine to reduce its redox potential and make it suitable for anaerobes.

The solid selective medium is Columbia agar, a highly nutritious medium which is frequently supplemented with a variety of nutrients and selective agents to render it suitable for different types of bacteria. In this case gentamicin is added at a concentration of $20 \, mg \, l^{-1}$ to restrict the growth of aerobic bacteria; clostridia, in common with other anaerobes, are resistant to gentamicin. Clostridia which grow on Columbia agar (with or without gentamicin) rarely display a characteristic appearance (Plate 30), so the presence of colonies which display a negative catalase reaction (fail to produce bubbles of oxygen on addition of a drop of 10 volume hydrogen peroxide) and consist of Gram-positive, rod-shaped bacterial cells, is regarded as a positive identification.

## 6.9  Summary

The final stages of the detection of the pharmacopoeial specified micro-organisms is crucially dependent upon the correct recognition of the organisms growing on solid selective media. Their appearance may differ from standard textbook descriptions of the species in question as a result of strain variation, variations in incubation temperature and time (even within pharmacopoeial limits) and subtle variations in culture medium composition. Furthermore, relatively harmless species which mimic the biochemical properties of the pharmacopeial-specified organisms are often encountered as contaminants of pharmaceutical materials, and these represent a source of confusion and may even lead to false-positive identifications. The Plates described in this chapter are intended as an aid to correct recognition of the organisms which are the subject of pharmacopoeial tests, and the reader may also find microbiology atlases (Koneman *et al.*, 1992) and illustrated textbooks (Varnam and Evans, 1996) useful in this respect.

## References

*British Pharmacopoeia* (1999) The Stationery Office, London.

*European Pharmacopoeia* (1997) 3rd edn. EP Secretariat, Strasburg.

HOLT, J.G., KRIEG, N.R., SNEATH, P.H.A., STALEY, J.T. and WILLIAMS, S.T. (1994) *Bergey's Manual of Determinative Bacteriology*, 9th edn. Williams & Wilkins, Baltimore, p. 151.

Koneman, E.W., Allen, S.D., Janda, W.M., Schreckenberger, P.C. and Winn, W.C., Jr (1992) *Colour Atlas and Textbook of Diagnostic Microbiology*, 4th edn. Lippincott, Philadelphia.

*United States Pharmacopoeia XXIII* (1995) United States Pharmacopoeial Convention, Rockville, MD.

Varnam, A.H. and Evans, M.G. (1996) *Foodborne Pathogens; an Illustrated Text*. Manson Publishing, London.

# 7

# Rapid Methods for Enumeration and Identification in Microbiology

PAUL NEWBY

*GlaxoWellcome Product Supply, Harmire Road, Barnard Castle, Co. Durham, United Kingdom*

## 7.1 The need for change

Two vital functions in pharmaceutical microbiology are the enumeration and identification of micro-organisms found in products and the manufacturing environment. The techniques currently used to carry out these tasks are firmly rooted in the past. Methods widely used in the industry today for microbial enumeration include pour plating, spread plating, most probable number (MPN) and the Miles and Misra technique (see Chapter 4). Commonly used methods of identification include visual inspection of colonial morphology, agglutination tests, selective differential agars and biochemical evaluation. Many of these methods would be familiar to the founding fathers of microbiology such as Pasteur, Koch and Petri who worked in the latter part of the 19th century.

For many decades these methods have sufficed because collectively they have the following advantages: the use of inexpensive materials; simplicity; the requirement for little specialist knowledge; their technically undemanding nature; and their acceptability to the regulatory authorities. However, a major drawback is the length of time taken to obtain a result, which can be from several days to several weeks depending on the type of test.

The pharmaceutical industry today is undergoing unprecedented change. The present situation is characterized by a prosperous, research-driven industry, but the whole economic structure, which in the past guaranteed high profits for the industry, is now changing fundamentally. Profit margins are being squeezed at the same time as development and manufacturing costs are increasing. Pressure is mounting for drug manufacturers to cut costs and reduce development and manufacturing lead times, particularly in the area of clinical trials. There is an absolute requirement in the pharmaceutical industry for research. Increasing research costs combined with high shareholder expectations dictate that time to market must be reduced to help maximize returns. Additional challenges include shorter patent life, an increasingly complex registration process and declining exclusivity – i.e. the time taken for competing products to appear – formerly it was years, it is now months. Increasing cost efficiencies will become paramount, and technology will play an important part in drug target selection, for example, combinatorial chemistry, high-throughput screening. Effective knowledge management and process optimization will be critical to these new technologies, and rapid microbiological methods and automation will increase handling capacity, will help reduce time delays and help in overall process optimization. Current microbiological test methods must change to reflect this. New methods which provide results quickly and which satisfy the exacting requirements of the pharmaceutical sector are needed.

## 7.2   Enumeration

### 7.2.1   *Current methods*

The pharmaceutical microbiologist must be able to enumerate micro-organisms present in product and production environments. Micro-organisms, by their very nature, are too small to be seen directly by the unaided human eye. Methods of counting have therefore relied either on direct observation using a microscope, or on microbial replication to produce visible colonies or turbidity. Direct microscopic enumeration is less commonly encountered than plating or MPN methods, most probably due to the laborious nature of this technique and also to the difficulty in determining cell viability.

   The basic unit of microbial enumeration is the visible colony forming unit, or c.f.u. The underlying assumption is that one viable micro-organism produces one c.f.u. A number of cell divisions are therefore needed to produce a visible colony, and this begins to explain why results using such methods take so long. It is not surprising then that new technologies have tried to move away from relying on microbial replication – a move which has not been entirely successful due to the requirement for single cell detection by the pharmaceutical microbiologist.

### 7.2.2   *New technologies*

New technologies are beginning to emerge which offer the promise of more rapid results, and most of these have their origins in the food and beverage sectors. Direct transfer from these sectors has not been easy due, in part, to the enormous differences in microbial specifications. Strong contenders which offer some hope to the pharmaceutical sector include: ATP bioluminescence; impedance; and chemiluminescence (Table 7.1).

### *ATP bioluminescence*

The firefly beetle, *Photinus pyralis*, produces light naturally in a process called bioluminescence. This reaction is mediated by the enzyme luciferin and requires the nucleotide adenosine triphosphate (ATP). The reaction was first investigated by McElroy (1947). ATP bioluminescence-based kits are today available commercially as quick and convenient ways of measuring microbial biomass. The technique is widely used in the food and beverage sectors, but with current detection limits of the order of 100 to 1000 c.f.u., lack of sensitivity has prevented the use of bioluminescence in the pharmaceutical industry. However, recent innovations and future developments are set to change this situation.

**Table 7.1** Suppliers of new technologies for pharmaceutical microbiology

| Type of technology | Supplier | Product |
|---|---|---|
| Chemiluminescence | Chemunex | Chemscan RDI |
| | | Digicount |
| ATP bioluminescence | Celsis Limited | RapiScreen |
| | | Celsis digital |
| | Hughes Whitlock Ltd | Bioprobe |
| | Millipore | SteriScreen |
| | | MicroCount Digital |
| Impedance | BioMerieux | Bactometer |
| | Don Whitley Scientific | Rabbit |
| | Malthus | Malthus System V |

*The reaction:* The enzyme luciferase catalyses the hydrolysis of ATP to produce yellow-green light (maximum emission at wavelength 562 nm) (Lundin, 1989). The reaction requires luciferin as substrate and is dependent on the presence of oxygen and magnesium. When all reagents are present in excess, the amount of light produced is proportional to the amount of ATP present. All living systems contain ATP; microbial cells contain relatively constant amounts at approximately $1.0–1.5 \times 10^{-15}$ g in a typical bacterial cell and $1.0–1.2 \times 10^{-14}$ g in a yeast cell (Lundin, 1989). Non-viable cells either do not contain ATP or else they lose the nucleotide rapidly. ATP bioluminescence is therefore a very useful indicator of microbial viability.

Essential steps in the detection of microbial ATP include: removal of non-microbial ATP; extraction of microbial ATP; light production, and light detection. The most commonly used extraction method found in commercial reagent kits utilizes cationic detergents. Once microbial ATP has been extracted it can react with commercially available luciferin/luciferase reagents, and the light produced is detected in a luminometer which normally incorporates a photomultiplier tube. The detection levels, which are in the region of $100–1000$ c.f.u. $ml^{-1}$, are influenced both by the low levels of ATP in microbial cells and by the quantum efficiency of the photomultiplier tube, which can be as low as 0.25% (Stanley, 1989).

*ATP bioluminescence in the pharmaceutical industry:* Single cell detection is required by the pharmaceutical microbiologist, so the current level of detection offered by ATP bioluminescence is clearly not sufficient. Thus, either ATP levels must first be amplified in some way, or alternatively the sensitivity of the detection system must be increased substantially.

Research is currently under way to help amplify the available pool of ATP and also to increase the sensitivity and robustness of the bioluminescent reagents. The use of alternative detection systems such as coupled charged device (CCD) cameras also offers the possibility of greater detection sensitivities. However, these methods are currently not available. A recent approach by the Millipore Corporation has been to combine traditional microbiological methods with ATP-based end-point detection systems. ATP levels are amplified by traditional enrichment procedures and viability is detected using bioluminescence. Two products have been launched aimed specifically at the pharmaceutical sector: the MicroCount digital system for water testing; and the SteriScreen system for non-sterile product screening.

With the MicroCount digital system, samples are filtered onto a compartmentalized filter plate through a specially designed sample chamber. The filter plate is incubated with nutrient medium for 24 – 48 h, after which viability is detected by bioluminescence. The system enumerates via a MPN-based approach (see Chapter 4). A 'memory' plate also provides an identification potential. Results are available after 1–2 days compared with 3–5 days using conventional methods. A recent report claimed a sensitivity range of between 1 and 76 c.f.u. in 100-ml samples (Wills *et al.*, 1997a).

The SteriScreen is a presence/absence test for non-sterile products. In practice, such products are for the most part, free from microbial contamination. The system, however, is not suitable for products with low bioburden present, even if the levels are within specification. Product is inoculated into broth and incubated for 24–48 h. The result, expressed in relative light units (RLU) is not quantitative due to this pre-enrichment step. Viability is determined by ATP bioluminescence. SteriScreen is suitable for a wide range of product types, including tablets and liquids, and can be used for preparations such as creams, ointments and suppositories, all of which are difficult to filter. It is important to determine the effect of any new product on the SteriScreen reaction. Product enhancement or inhibition of the luciferase-mediated reaction must be shown to be minimal. In addition, the product must not contain high levels of residual ATP, which might interfere in the reaction and cause false-positive results. This system offers the possibility of reducing the Microbial Limit Test for non-sterile product release, from 5–7 days down to 1–2 days, a reduction which represents a significant saving in time. SteriScreen has been successfully validated at a number of pharmaceutical manufacturing sites for a range of non-sterile formulations including syrups, tablets and suppositories (Wills *et al.*, 1997b). In the United Kingdom, regulatory approval by The Medicines Control Agency has recently been granted for use of SteriScreen for a non-sterile pharmaceutical formulation (Weatherhead, 1997).

An alternative system is the Hughes Whitlock Bioprobe (Whitlock, 1994). This is a very versatile, portable, hand-held luminometer system with which it is possible to detect ATP from swabs, liquid samples, direct surfaces, biofilm disks and from filter membranes. The basic unit is adapted using a range of different sample devices. Using the filtration option, for example, the sample is filtered directly onto a standard filter membrane. Extractant and bioluminescent reagents are spread across the surface of the membrane, and the Bioprobe is used to detect light output. Sample volumes can be increased to improve the detection sensitivity of the system. Current reagents offer sensitivities in the range of 100–1000 c.f.u., but new reagents are under investigation which will increase detection sensitivities further. Potential uses for the Bioprobe could include water testing, non-critical environmental monitoring, in-process bioburden and pre-filtration bioburden testing.

## Impedance

Impedance – which is defined as the resistance to the flow of an alternating current through a conducting material – can be used to detect and measure microbial growth. This idea is not new, having first been suggested 100 years ago (Stewart, 1899). During the past two decades, impedance-based systems have become available and are widely used in the food and beverage industries.

Impedance ($Z$) can be defined in terms of resistance ($R$) and capacitance ($C$) by the equation:

$$Z = \sqrt{R^2 + (\tfrac{1}{2}\pi F C)^2}$$

where F equals the frequency (Silley and Forsythe, 1996). A reduction in conductance (reciprocal of resistance) or capacitance leads to an increase in impedance. As micro-organisms grow, they metabolize complex constituents of the medium, producing smaller more highly charged products such as acids, carbon dioxide and amino acids which alter the electrical properties of the growth medium. As the micro-organisms multiply, a *detection threshold* is reached where an electrical signal is generated; generally this level is approximately $10^6$ c.f.u. ml$^{-1}$ (Silley and Forsythe, 1996). The time taken from initial inoculation to signal detection is called the *detection time*; this is inversely proportional to the size of the initial inoculum, and will vary for different types of media and for different types of organisms.

*Direct impedance:*   This is the measurement of the passage of an electrical current through a conduction medium. The efficiency of this process relies on the type of medium and organism under investigation (Owens, 1985). Direct impedance is less effective with buffered media and some selective media. Yeast and moulds do not produce large changes in impedance as they do not produce strongly ionizing metabolites.

*Indirect impedance:*   This relies on the ability of the test organism to produce carbon dioxide. Electrodes are placed in a carbon dioxide sink such as potassium hydroxide, and changes in impedance result from the ionization of the carbon dioxide (Owens *et al.*, 1989). This system has been shown to be useful for a range of micro-organisms including *Staphylococcus aureus*, *Escherichia coli*, *Bacillus subtilis* and *Pseudomonas aeruginosa* (Bolton, 1990). Indirect impedance would allow the use of compendial media and selective differential media unsuited to direct impedance.

*Applications of impedance in the pharmaceutical industry:*   Impedance-based test systems are used by the food industry for raw material and end-product testing, but they are not widely used in the pharmaceutical sector. This is due partly to the difficulty in using compendial media in direct impedance systems, thus raising regulatory compliance concerns. There has also been a reluctance in the pharmaceutical industry to use indirect impedance-based systems, due perhaps both to ignorance of the technique and to an innate conservatism. An additional factor is the relatively high price of the systems given their limited areas of application.

Preservative efficacy screening is one area of successful application of impedance. Studies carried out by Connolly *et al.* (1994) showed very favourable results using a range of preservatives for *Staphylococcus aureus*, *Candida albicans*, *Aspergillus niger* and *Pseudomonas aeruginosa*. By using impedance for the initial screening, regulatory issues are avoided. A compendial test can be used to confirm initial results. Moreover, considerable savings in sample preparation time and resource are possible with impedance-based systems, with preservative screening of optical solutions having been reported (Miller, 1997). Although interest in using impedance for sterility testing has been expressed (Dal Maso, 1998), a fundamental re-design of the test system will be required in order to maintain sterility during incubation. The regulatory considerations are considerable, particularly in view of the trend for increased incubation times.

## Chemiluminescence

Chemiluminescence is the process of producing light with the aid of a chemical fluorophore. Two chemiluminescence-based techniques of interest for the enumeration of micro-organisms are direct epifluorescent technique (DEFT) and Chemscan, manufactured by Chemunex.

*DEFT:*   This method was first developed for the rapid analysis of raw milk (Pettipher and Rodrigues, 1981) and subsequently used for a range of applications in the food industry. Samples are filtered and micro-organisms stained with the viability indicator acridine orange; viable organisms appear orange and non-viable organisms green when viewed under epifluorescent microscopy. Unfortunately, heat treatment of samples above 50°C can invalidate the differential properties of acridine orange, and this has therefore limited its use in the pharmaceutical sector. However, some investigations have been made using micro-colony formation to assess pharmaceutical grade waters (Denyer and Ward, 1983; Newby, 1991). The sensitivity of the technique has been hampered by the inability of available systems to differentiate between fluorescing debris and true micro-organisms, in addition to the consequent labour-intensive microscopic examination required by the operator.

*Chemscan:*   In many respects the Chemscan follows on from DEFT. It may offer the pharmaceutical industry the exciting advantages that DEFT gave the dairy industry by providing highly sensitive detection in near-real time. Like DEFT, Chemscan is based on microbial capture by filtration and subsequent fluorescent labelling of cells with a viability indicator. Where it differs from DEFT is in label detection by laser scanning. Key features of the Chemscan are: the viability indicator is a derivative of carboxy fluorescein; fully automated image analysis; the potential of single cell detection; and high discrimination to eliminate false positives due to particles.

Central to the whole Chemscan process is the viability indicator Chem-Chrome, a fluorescein derivative which is non-fluorescent. During staining, cells take up Chem-Chrome and cleave it by intracellular esterase activity to produce the fluorochrome fluorescein (Wallner *et al.*, 1997). In addition, only cells which have an intact cell membrane can retain fluorescein in sufficient quantities for subsequent detection by laser scanning. Sophisticated software processes the data generated by the laser scan, giving the Chemscan the ability to differentiate automatically between fluorescing micro-organisms and fluorescing debris. This represents a significant advance over DEFT and makes for much faster and less laborious sample processing.

*Uses of Chemscan in the pharmaceutical industry:*   Chemscan is an emerging technology, but has already been identified as a rapid method for the analysis of pharmaceutical grade waters (Wallner *et al.*, 1997). Results are available for both vegetative organisms and spore-formers (using an accelerated germination process) within 2 h, thus allowing rapid corrective action to be taken. Chemscan is ideally suited for troubleshooting in water systems where rapid results would not only assist greatly in any attempt to sanitize the system, but also would rapidly identify if further corrective action was necessary. A multi-site evaluation is under way involving a number of major pharmaceutical manufacturers and Chemunex, manufacturers of the Chemscan RDI. The aim of this study is to carry out independent validation of the system and also to gain regulatory acceptance of the method.

Another area of potential use of this system is the detection of viable non-culturable (VNC) organisms, particularly, but not exclusively, in water systems. Microbiologists have recognized that many different types of bacteria cannot be cultured or can enter into periods of non-culturability (Kell *et al.*, 1998; McDougald *et al.*, 1998). Results from water analysis indicate greater sensitivity of the Chemscan compared with conventional culture techniques (Wallner *et al.*, 1997). Regulatory authorities are considering the impact of new technologies, and for water analysis the *European Pharmacopoeia* (1997) monographs are including statements such as 'where direct counting techniques are used, other limits may need to be applied to take increased recovery into account' (Hargreaves, 1999).

Samples must be filterable for use with the Chemscan. With this limitation in mind there are a number of areas of potential use for the instrument: raw material testing (including water); in-process testing; non-sterile product testing; and preservative efficacy screening.

The manufacturers of the Chemscan RDI have recently introduced a new product, the 'D-Count' for the testing of non-filterable non-sterile products. D-Count is a flow cytometer-based system which uses digital electronics to detect and enumerate viable micro-organisms. The system incorporates the Chemunex cell labelling system, laser excitation and digital data processing technology already used for the ChemScan RDI. Micro-organisms are labelled using the viability indicator Fluorassure. After labelling, the cells pass individually through a laser excitation beam where a sensitive photo-multiplier is used for detection. Discrimination parameters are used to differentiate between labelled, viable organisms and autofluorescing particles. The current sensitivity of the D-Count is in the region of $50\,c.f.u.\,ml^{-1}$. Results are expressed as direct counts per volume and can be stored on a personal computer for traceability. The system is currently being used to test a range of complex sample matrices such as toothpaste and ointments.

Currently, the Chemscan can enumerate micro-organisms but has no identification capacity. Consequently, a research programme is under way to develop a fluorescent antibody-based detection system, and by using such an approach it should be possible to demonstrate the absence of specific organisms from water and pharmaceutical products. This approach could not readily be applied to the identification of unknown isolates, but may be an important development for non-sterile product testing where the Microbial Limit Test (MLT) is used for product release. This consists of an enumeration test and an examination for the absence of pathogens (see Chapter 6). Thus, with a rapid pathogen identification and total enumeration capacity combined, Chemscan would offer the potential for same-day product release. This is an exciting development and one that would bring real-time microbiological testing much closer.

## 7.3 Identification

### 7.3.1 *Current methods*

Identification methods used by most pharmaceutical microbiologists today still usually employ microbiological culture. Current methods rely principally on the separation of mixed populations into pure cultures, and this is achieved using: selective medium enrichments and agars which actively select for specific species or groups of micro-organisms; purity plates which demonstrate pure culture has been achieved; and serial dilution separation techniques. Identification then relies on variously, colonial morphology, selective growth, biochemical tests and agglutination tests (see Chapter 5). Immunological methods such as enzyme-linked immunosorbent assay (ELISA) and other fluorescent antibody techniques have not been widely adopted.

Identification in pharmaceutical microbiology is required for a range of purposes associated with quality assurance: determination of organisms from the manufacturing environment; identification of contaminants from final product testing; demonstrating absence of named organisms from non-sterile products and water; quality control of fermentation stocks in biotechnology; and confirmation of test organisms in validation processes.

### 7.3.2 *Automated identification systems*

Regulatory authorities increasingly require already busy laboratories to extend the amount of identification carried out on an organism isolated from the manufacturing environment. In order to meet this demand there is a need for faster, more accurate techniques. The impact of taxonomic changes creates confusion and is evidence that current identification methods are too insensitive to separate closely related species with any accuracy. For these, and other reasons, there has been an enthusiastic welcome across the industry for automated microbiological identification (AMI) systems. This is in marked contrast to the introduction of automated enumeration techniques. The most widely used AMIs rely on microbial specificity for specific substrates. Examples of substrate-utilizing AMIs include the BioMerieux Vitek and the Biolog system. An alternative method to substrate utilizing AMIs involves fatty acid analysis by gas chromatography.

### *Vitek*

The Vitek system was developed in the 1960s for the identification of micro-organisms in the urine of astronauts in space (Stager and Davis, 1992). Identification is dependent on the reactions produced from the metabolism of a number of biochemical substrates. Vitek is one of the most widely used automated systems in the pharmaceutical industry, and comprises: 30-well identification cards; a filler/sealer unit; a reader/incubator unit; a data terminal and printer.

Sample preparation is simple, and involves subculturing isolates to obtain pure cultures followed by Gram staining, oxidase, catalase or coagulase testing and suspension in saline. The wells of the appropriate identification card are then vacuum-filled with saline suspension, and the filled cards are inserted into the reader/incubator. Up to 480 cards can be analysed, the cards being read at hourly intervals by sensors looking for changes in turbidity or colour; the reaction pattern is analysed by computer. Information is compared with a principal database called the *primary react file* (PRF), and identification is made on the basis of this comparison. If a result is not possible using the PRF, a second database – the *secondary react file* (SRF) – can be used. Selection parameters are relaxed in the SRF compared with the PRF, and this must be taken into consideration when reviewing any result. The SRF is useful for the identification of industrial environmental isolates, but the system also offers database management, including trend analysis of test results.

Identification by Vitek is obviously limited by the range of cards available. This is particularly true for environmental isolates, though new cards are currently under development by the manufacturer. Although the SRF can help customize the user's database, identification of some obligate aerobes is a problem with the Vitek due to the vacuum required for card filling which prevents sufficient growth for identification to occur. Vitek is not suitable for the identification of *Micrococcus* species. It is critical that the Gram stain reaction is correct; if not, the wrong identification card will be used, leading to erroneous results. Results from the Vitek must be read in context with the preliminary Gram stain result and the environment from which the sample was originally taken, since the Vitek (or any other automated identification system) is not a substitute for a properly trained microbiologist.

In summary, Vitek is widely used across the pharmaceutical industry as it provides rapid, straightforward and objective identification, has a large sample handling capacity, and can offer considerable labour savings in sample preparation, data analysis and reporting. Vitek can provide results in the majority of cases between 8 to 15 h, which is similar to the Biolog (see below). API results take perhaps slightly longer, from 15 to 28 h, while conventional selective agars can take anything between 18 to 48 h or more to provide a result.

## Biolog

In the Biolog, carbon utilization produces colour changes describing a metabolic fingerprint, the identification being made by comparison with an extensive database. The Biolog is available as either a manual or automated system. The automated system consists of a manual eight-channel repeating pipettor, a turbidimeter, a microplate reader and a MicroLog program disk. An IBM-compatible computer is also required for data processing.

A microbial saline suspension is prepared and inoculated into individual wells of a microplate containing a variety of carbon substrates. The plate is incubated at 30–35°C, with readings being taken after a minimum of 4 h, either manually or by using the microplate reader. For bacteria, carbon-source oxidation is detected when increased cell respiration leads to irreversible reduction of a tetrazolium dye, resulting in a purple colour change. For yeasts, the test plate contains both assimilation and oxidation tests. The resulting colour patterns must be analysed by software, and identification given after comparison with the database which encompasses most known human pathogens and major industrial environmental isolates. In general, bacterial results are obtained between 4 and 18 h. Advantages of the Biolog include easy sample preparation and handling, objective identification and manual or automated sample reading facilities; most importantly, the database includes a wide range of clinical and environmental isolates. Again, like Vitek, it is essential to use pure culture preparations and the correct Gram stain result.

## Gas chromatography

Gas chromatography (GC), offers an alternative to substrate-utilizing identification systems. Identification follows the extraction of cellular fatty acids which are analysed by GC and compared with a central database. Fatty acid composition is a useful tool for identification due to its highly conserved nature within a taxonomic group (Stager and Davis, 1992). The MIDI Microbial Identification System (MIS) (Microbial ID, Inc., Newark, DE, USA) is an automated GC-based identification system which can analyse more than 300 fatty acid methyl esters ranging in length from nine to 20 carbons.

GC analysis requires trained operators and involves the use of complex and sensitive equipment. Moreover, culture conditions need to be carefully controlled in order to provide reproducible fatty acid profiles. Solvent extraction of cellular fatty acids is also a lengthy and multi-stage procedure. The software system allows the user to generate databases of frequently encountered isolates, and this is particularly useful when isolates are difficult to identify. Cluster analysis of data is also possible which generates dendograms or two-dimensional plots for comparative analysis of isolates – a very useful method for tracking the source of contamination during troubleshooting investigations.

### 7.3.3  *Future trends in identification*

Undoubtedly, genetic-based systems of identification will become increasingly important in the future, but they are not yet a reality. Automated systems are becoming available that simplify the whole process, although the cost of these units is high.

### Nucleic acid probes

Nucleic acid probes can be either RNA or DNA, and they can be used to identify specific bacteria, yeasts, moulds, viruses and protozoa. DNA probes consist of single-stranded DNA that binds to a specific region of the denatured target DNA. The double-stranded hybrid formed is then detected by some form of label. Gene probes are highly specific, but may require amplification of the target. By combining gene probes with the polymerase chain reaction (PCR), Mullis (1987) and Mullis *et al.* (1987) have shown that highly selective and sensitive identification is possible. One of the considerable advantages offered by gene probe technology will be sensitivities at the picogram level. Moreover, high specificity will provide greater taxonomic accuracy. The technology may offer a quick and relatively simple method for the routine detection and identification of viruses and other difficult-to-cultivate organisms, and may allow the detection of foreign organisms in a bioreactor.

However, there are also disadvantages to this technology, as routine laboratories will need to invest in new equipment and staff training. Interference in the assay from formulation excipients, contaminants and non-viable organisms is a very real concern; such interference could lead to confusion and false reporting. Furthermore, the method could not be used for sterility testing because non-viable organisms will be detected.

The revolution in identification using gene probes has already started, with commercial suppliers developing nucleic acid probe-based kits for the rapid identification of human immunodeficiency virus, hepatitis B and C viruses and cytomegalovirus, and for slow-growing organisms such as *Mycobacterium tuberculosis*. In addition, automated genetic identification systems are beginning to appear that greatly reduce the skill needed to use this technology. One such instrument is the RiboPrinter™ (a trademark of Qualicom, Inc., a subsidiary of E.I. du Pont de Nemours and Company).

The RiboPrinter's sample preparation is very simple in that it requires a pure culture grown on standard agars. A heat treatment step is followed by a restriction enzyme step;

*Eco*R1 is used to cut target bacterial DNA. A labelled rRNA operon probe hybridizes with the target denatured restriction fragments. The resulting pattern, or RiboPrint, is analysed and compared with a database. The RiboPrinter has been designed to perform most of these processes inside the unit, such that operator involvement is reduced to a minimum. Comparative analysis with the database allows the system to be characterized as alike or unlike, and finds application for the fingerprinting of organisms not already characterized (Bruce, 1996). This would be a very interesting facility for sterility test failure investigations, as any contaminant could be accurately compared with operator and environmental flora for clues as to the origin.

The commercial interest in nucleic acid probe technology is very keen, and the potential market for rapid, sensitive test systems is huge. The new genetic-based technologies have much to offer microbiology in general, and the high sensitivity and specificity of these systems is ideally suited to the needs of the pharmaceutical industry. Indeed, these techniques are set to revolutionize the way in which microbiological testing is performed in the future.

## 7.4   Conclusions

The need for rapid enumeration and identification techniques in the pharmaceutical sector has never been more acute. Many rapid enumeration methods taken up enthusiastically by other industries have failed to make an impact in pharmaceutical microbiology, due in part to the highly regulated environment and in part also to the failure of the manufacturers of these methods seriously to address the specific requirements of this sector. This situation is now changing, with dialogue increasing between the technology suppliers, potential users and the regulatory authorities, and methods are beginning to emerge and regulatory acceptance sought. Regulators wish to protect public health, without discouraging innovation. Approval was given by the MCA in 1997 for an ATP bioluminescence-based method for non-sterile product testing, while manufacturers such as Millipore and Celsis Lumac Ltd are providing information and seminars which are open to all, and Chemunex is providing seminars for inspectors. Regulatory approval is being actively sought for product testing with the ChemScan and D-count. The message here is that the regulatory authorities are receptive to new technologies, though an established regulatory 'road map' to aid the introduction of new technology has yet to emerge.

The situation for automated identification systems is somewhat different, in that there has been wide acceptance of such systems across the pharmaceutical sector for some time, and in many respects the pharmaceutical sector is ahead of others such as the food and water industries. The advent of rapid, convenient and accessible genetic-based techniques is set to improve identification practices across the whole of industrial microbiology. In addition, major changes occurring across pharmaceutical microbiology are bringing the prospect of real-time analysis closer to realization.

## References

BOLTON, F.J. (1990) An investigation of indirect conductimetry for detection of some food-borne bacteria. *Journal of Applied Bacteriology*, **69**, 655–61.

BRUCE, J. (1996) Automated system rapidly identifies and characterises micro-organisms in food. *Food Technology*, January, 77–81.

CONNOLLY, P., BLOOMFIELD, S.F. and DENYER, S.P. (1994) The use of impedance for preservative efficacy testing of pharmaceuticals and cosmetic products. *Journal of Applied Bacteriology*, **76**, 68–74.

DAL MASO, G. (1998) The use of the bioMerieux bactometer for sterility testing of pharmaceutical products. *Annali Di Microbiologica ed Enzymologia*, **48**, R7–R13.

DENYER, S.P. and WARD, K.H. (1983) A rapid method for the detection of bacterial contaminants in intravenous fluids using membrane filtration and epifluorescence microscopy. *Journal of Parenteral Science and Technology*, **37**, 156–8.

HARGREAVES, P. (1999) Regulatory issues and implications for the pharmaceutical industry. An oral presentation at the Pharmig, Parenteral Society and Royal Pharmaceutical Society Joint Seminar. *Viable but non-culturable micro-organisms – a live issue*. London, January 28, 1999.

KELL, D.B., KAPRELYANTS, A.S., WEICHART, D.H., HARWOOD, C.R. and BARER, M.R. (1998) Viability and activity in readily culturable bacteria: a review and discussion of the practical issues. *Antonie van Leeuwenhoek*, **73**, 169–87.

LUNDIN, A. (1989) ATP assays in routine microbiology. In: STANLEY, P.E., MCCARTHY, B.J. and SMITHER, R. (eds), *ATP Bioluminescence*. Blackwell, Oxford, pp. 11–30.

MCDOUGALD, D., RICE, S.A., WEICHART, D. and KJELLEBERG, S. (1998) Nonculturability: adaptation or debilitation? *FEMS Microbiology Ecology*, **25**, 1–9.

MCELROY, W.D. (1947) The energy source for bioluminescence in an isolated system. *Proceedings of the National Academy of Science USA*, **33**, 342–5.

MILLER, M.J. (1997) 'Use of the bioMerieux Bactometer system for screening new disinfectant and preservative formulations'. Presented at the PDA Annual Meeting, Philadelphia, November.

MULLIS, K.B. (1987) US Patent 4,683,22.

MULLIS, K.B., ERLICH H.A., ARNHEIM, N., HORN, G.T., SAILI, R.K. and SHARF S.J. (1987) US Patent 4,683,202.

NEWBY, P.J. (1991) Analysis of high-quality pharmaceutical grade water by a direct epifluorescent filter technique microcolony method. *Letters in Applied Microbiology*, **13**, 291–3.

OWENS, J.D. (1985) Formulation of culture media for conductimetric assays: theoretical considerations. *Journal of General Microbiology*, **131**, 3055–76.

OWENS, J.D., THOMAS, D.S., THOMPSON, P.S. and TIMMERMAN, J.W. (1989) Indirect conductimetry: a novel approach to the conductimetric enumeration of microbial populations. *Letters in Applied Microbiology*, **9**, 245–9.

PETTIPHER, G.L. and RODRIGUES, U.M. (1981) Rapid enumeration of bacteria in heat-treated milk products using membrane filtration-epifluorescence microscopy technique. *Journal of Applied Bacteriology*, **50**, 157–66.

SILLEY, P. and FORSYTHE, S. (1996) Impedance microbiology – a rapid change for microbiologists. *Journal of Applied Bacteriology*, **80**, 233–43.

STAGER, C.E. and DAVIS, J.R. (1992) Automated systems for identification of micro-organisms. *Clinical and Microbiological Reviews*, July, 302–27.

STANLEY, P.E. (1989) A concise beginner's guide to rapid microbiology using adenosine triphosphate (ATP) and luminescence. In: STANLEY, P.E., MCCARTHY, B.J. and SMITHER, R. (eds), *ATP Bioluminescence*. Blackwell, Oxford, pp. 1–10.

STEWART, G.N. (1899) The changes produced by the growth of bacteria in the molecular concentration and electrical conductivity of culture media. *Journal of Experimental Medicine*, **4**, 235–43.

WALLNER, G., TILLMAN, D., HABERER, K., CORNET, P. and DROCOURT, J.L. (1997) The Chemscan system: a new method for rapid microbiological testing of water. *European Journal of Parenteral Science*, **2**, 123–6.

WEATHERHEAD, M. (1997) 'Improved customer service provided by QC laboratory following introduction of new rapid test method for non-sterile products'. Presented at the PDA Annual Meeting, Philadelphia, November.

WILLS, K., FERGUSON, D., WOODS, H. and GRANT, P. (1997a) 'Multi-centre validation of rapid bioluminescent technique for the enumeration of micro-organisms in pharmaceutical process water' 97th ASM General Meeting, Miami Beach.

WILLS, K., TURTON, A., PICKETT M., BACCARINI, P. and PRESENTE, E. (1997b) 'Multi-centre validation of rapid bioluminescent technique for microbiological testing of non-sterile pharmaceutical end-products'. 97th ASM General Meeting, Miami Beach.

WHITLOCK, G.D. (1994) 'A method of directly measuring bacteria adhering to surfaces without removing them'. PITCON '94 – Instrument Development – Paper 1031.

**8**

# Microbiology Laboratory Methods in Support of the Sterility Assurance System

ANDREW BILL*

*Medicines Control Agency, 2nd Floor, Prudential House, 28–40 Blossom Street, York, YO24 1GJ, United Kingdom*

## 8.1  Introduction

This chapter is divided into the following main sections: sterility testing; bioburden testing; biological indicators; and environmental monitoring. The section on sterility testing includes many of the basic principles applicable to running a microbiology laboratory that deal with the testing in support of the sterility assurance system. The objective in all sections is to provide practical advice, and to encourage microbiologists to think all their activities through to a proper conclusion.

There are several references to European Normalization (EN) standards. These have been developed for medical devices, and those in the pharmaceutical industry may not be familiar with them. It is recommended that reference is made to them as they contain a wealth of information relevant to all activities connected with sterility assurance.

## 8.2  Sterility testing

### 8.2.1  *Preliminary considerations*

Sterility is a state of absence of living organisms and as such cannot be positively tested, i.e. it cannot be concluded from the results that the batch from which samples were drawn, or even the actual samples tested, were sterile. This uncertainty is because there may be organisms present that cannot grow and reveal themselves under the culture conditions used, or there may be insufficient in number for any to be effectively sampled. At best a sterility test detects growth in broth which may or may not prove that the batch sampled is non-sterile. Nevertheless, the presence of growth can have considerable influence on whether or not the batch is released for sale. The sterility test provides an opportunity to detect batches that should not be

---

* The content of this chapter should not be taken as the policy of the Medicines Control Agency or the Medical Devices Agency.

released, and as a consequence reveals processes that may be out of control. The result of a sterility test provides little evidence on its own to justify release. The same logic can be applied to other microbiological tests used in critical operations such as aseptic processing, where the absence of positive units in a broth fill or absence of growth on a finger dab plate provides little confirmation that everything is in control, whereas presence of growth provides valuable information about loss of control.

The term parametric release is used to describe batch release without sterility testing of product in the pharmaceutical industry. In the UK sterile medical devices industry, release of batches without sterility testing of product is normal and the term is applied to the release of batches from ethylene oxide sterilization processes without testing the process biological indicators (BIs). For medical devices the requirements for this type of parametric release are defined in relevant European standards (EN 550, 1994). Where sterile medical devices are steam-sterilized or sterilized by irradiation, sterility testing of product may or may not be carried out depending upon country of destination, type of device and company practice.

Sterile UK pharmaceuticals are sterility tested if they are aseptically prepared, and nearly all terminally sterilized products are sterility tested. However, the mechanism for approval for batch release of terminally sterilized pharmaceuticals without sterility testing is being considered by European regulatory bodies. Any system for reducing or deleting sterility testing will inevitably require the use of best practice and for the company to demonstrate a rigorous control of pre-sterilization count, sterilization processing and physical separation to prevent unsterilized product being packed and released. Design of the product should clearly be such that there are no questions about product integrity following the sterilization process.

US parametric release also focuses on the use of biological indicators or sophisticated thermochemical indicators. In Europe there are no prospects for waiving the need for sterility testing of aseptically made product because of the intrinsic lack of assurance of such processes when compared with conventional terminal sterilization.

### 8.2.2 *Application*

Sterility testing is carried out to support batch release of sterile pharmaceuticals and, if applicable, medical devices using pharmacopoeial methods or variants. Versions of the sterility test are also carried out as retrospective monitoring of sterile pharmaceuticals prepared for immediate use under a 'specials' manufacturing licence in the UK. A type of sterility test is also used to support irradiation dose setting (EN 552, 1994). Another application of the test is as part of the validation and re-validation of sterilization processes, particularly those where routine product sterility testing is not intended. Other applications, usually in the full reference pharmacopoeial form, include testing of recalled or retained samples as part of an investigation into complaints or incidents and to support shelf-life determinations.

### 8.2.3 *Sampling*

Pharmacopoeial sampling rates are the starting point for devising sampling programmes with due regard being given to sampling worst case or units at highest risk (see Chapter 3). Highest risk in aseptic processing may be at start-up, during interventions, after stoppages, during changes of bulk supply, and at the end of the filling run. For terminal sterilization, units would be chosen from positions where extremes of the process were noted during validation. In all cases sufficient additional units should be taken to be representative of the whole process.

Samples should be taken after standard checks for container integrity, i.e. they should represent what would be sold. Once removed from the batch, the unused samples should not be returned (because they may be returned to the wrong batch); re-test samples should be identified and stored taking this into account. Clear labelling of samples is essential; the worst situation is that an unfit batch is released due to sample mislabelling. Sample storage conditions should mimic normal handling of the product but, bearing in mind that the units will subsequently be tested, measures should be taken to limit build up of micro-organisms on surfaces that may be handled during aseptic manipulation for sterility testing. Time taken in handling samples properly is usually under-estimated. Mislabelling, mishandling and loss of samples can result in errors and poor quality decisions being taken to recover the situation and justify batch release.

The provision of special test packs for particular applications of sterility testing is sometimes unavoidable. They are of particular relevance to complex or large medical devices where disassembly is essential to enable the exposure of critical surfaces to nutrient broth. Aseptic disassembly may add so many handling contamination risks that the sterility test becomes meaningless, so a partially disassembled device is packed and sterilized with the batch as a special unit for subsequent sterility testing. Such packs may not represent the standard product exposure during the sterilization process, so alternatives such as using high-grade isolators for product disassembly are better options.

### 8.2.4  *Facility and equipment*

The interpretation of results from sterility testing can be confused by the risk of microbiological contamination entering during the test and masking a genuine non-sterile unit or falsely failing a sterile unit. Conventional clean room facilities used for sterility testing may vary from a horizontal laminar flow hood in a dedicated room with minimal laboratory coat change, to a full aseptic suite with sterile gowning and grade B background as described in Annex 1 to the Guide to Good Manufacturing Practice for Medicinal Products [Rules and Guidance for Pharmaceutical Manufacturers and Distributors ('Orange Guide') 1997].

These facilities tend to generate a positive broth rate of 0.1% up to 5%. The complexity of testing, skill of people, surface sanitization methods and exactly how the positive growth rate is calculated affect the positive frequencies seen, and of course some of these may be due to genuinely non-sterile product rather than contamination during testing. Secure aseptic techniques for sterility testing can be devised using horizontal flow hoods; vertical (down-flow) laminar airflow hoods are considered unsuitable without special equipment and methods, owing to interference during operations with the flow of filtered air.

Isolators are increasingly being used for sterility testing. The highest grade versions of these isolators employ high-quality construction materials, rigorous methods for leak testing, and effective surface decontamination using peracetic acid, hydrogen peroxide or formaldehyde gassing. Such isolators are returning contamination rates of zero, or very nearly so, and thus seem to be the solution to the false-positive problem. Given that such technology exists, the rationale for continuing to use lower-grade facilities if they persistently return high false-positive rates is difficult to condone.

In all types of facility, the basic quality system needs to ensure that appropriate maintenance, routine checking and periodic calibrations are carried out. Lack of attention to these fundamental precautions can affect positive rates. The provision of basic items of equipment, for example vacuum reservoir bottles, spray bottles, Bunsen burner lighters, tube clamps, gas hoses, chairs and their replacement at the first sign of damage, wear or lack of efficiency yields handsomely in terms of operator performance and reduction in false positives. The culture of excessive financial management and 'making do' is false economy in sterility test facilities because of the penalty in terms of delay in batch release or batch discard arising from false positives. Finally, the facility ought to be designed based on ergonomic principles, e.g. the operator working position should be comfortable, sufficient lighting should be available, and distractions while carrying out critical techniques are prevented.

### 8.2.5 *People*

Sterility testing is best conducted under the supervision of a qualified microbiologist or a person qualified by relevant experience and proven performance. People are largely the source of variation in false-positive rates from conventional clean room facilities. In addition to person-to-person variation, a poor emotional state in a top class operator can result in a run of false positives. In common with all manual aseptic operations, the operator needs to carry out complex manipulations respecting the direction of filtered airflow, so manual dexterity, visual acuity and ability to sustain concentration are absolute pre-requisites. Given these natural abilities, the design of the facility should enhance their expression, and hours of working and working pattern should be considered as major factors, for example standard daytime work with perhaps a 2-h limit for a continuous work session would be ideal.

Training should include, as a minimum, the following elements: sample management; documentation; incubator management; examination and recording of broth results. Training programmes should include: clear descriptions of the purpose of testing and how the results will be used; practical demonstrations; assessment of comprehension of written procedures; dummy runs on the bench to become familiar with handling equipment; practice under supervision; formal evaluation, and when competence is achieved, a sign-off by trainer and trainee.

Aseptic manipulations require greater attention. In addition to the steps above, the individual capabilities of each operator need to be recognized. It is presumed that the optimum aseptic technique has been developed for every operation. It is not enough to give an instruction to surface sanitize with 70% isopropyl alcohol, since each item to be sanitized will have a defined best way identified; similarly the process of, for example, pouring an aliquot or making a closed system aseptic connection should be defined in every detail. Where the operator cannot carry out the prescribed process, appropriate approved modification can be made, but the system should not be compromised and there may come a point where the person is deemed to be unsuitable for that task.

Once the operator has demonstrated basic capability, as many dummy runs as necessary are needed to build up to the correct speed of operation. This does not relate to productivity, but to minimizing the risk of accidental airborne contamination by reducing the exposure time of surfaces at risk, consistent with preventing touch contamination. Once this stage is reached the operator can carry out sterility tests on sterile product designated for such trials where positive results can be attributed to operator training and not recorded formally on the batch records with consequent implications for company integrity. Having completed the designated number of trials without contamination occurring, the operator is ready for formal certification for each type of sterility testing carried out in the facility. This may be accomplished by sterility testing actual product for say three batches under decreasing supervision with a sign-off based on satisfactory results. When stand-in operators need to refresh sterility test skills or interruptions in service take place, a revision and re-certification process should be devised. Aseptic procedures should be audited periodically (perhaps twice a year) to ensure that optimum methods are still being applied.

Where isolators are being used, standards should still be maintained, even though with gassed systems and positive pressures, environmental contaminants are not expected. The difficulty in working with thicker gloves in an isolator means that the same high level of visual acuity and sustained concentration is necessary to compensate for a slight loss in manual dexterity and the additional need to protect the isolator envelope from damage from needles, etc. Also, although environmental contaminants are not expected, there is always a slight possibility that they may be present; aseptic techniques can readily reduce this risk.

### 8.2.6 *Methods*

The pharmacopoeias are specific about test methods and when a pharmaceutical product licence reflects sterility tests according to *British Pharmacopoeia, European Pharmacopoeia* and *United States Pharmacopoeia* (BP, EP, USP) the use of the specified methods is the easiest way to comply. If a variation in some detail of the pharmacopoeial method is used, then equivalence should be established. The published methods basically rely on: use of specified microbiological broths to bathe the surfaces of containers declared as sterile, or immersion of items in broth, or mixing material with broth or retention of the micro-organisms on a filter and culture of the filter in broth. The filtration methods are most commonly used for pharmaceuticals, and the immersion or filling methods are more applicable to medical devices.

One of the key points is to demonstrate that the item, or material being tested, does not inhibit the outgrowth of any micro-organisms that might be present which might cause a false pass decision. The presence of inhibitory materials is tested for as described in the specified methods. It is still incumbent on those responsible for sterility testing to observe the spirit of the requirements and to provide properly for overcoming any antimicrobial property. This may arise when it is known that the standard test organisms would grow, but other types may not. Also if the process or plant has an abnormal, but consistent flora not represented by standard test organisms, their inclusion in the challenge testing is sensible.

In practice, the sterility test usually involves the following stages: preparation of equipment, rinse fluids and sanitization fluids; preparation of the workstation and transfer of items to it; filling, immersion or filtration and addition of broth; incubation; and examination, recording results and sentencing.

### 8.2.7  *Preparation and materials*

Surfaces of samples exposed in the workstation should be rendered visibly clean and, if known to be at risk from microbial contamination, may be subject to a preliminary decontamination step (see section 8.2.11 for filter holders). Decontamination may be as simple as a detergent wash, followed by a rinse with sterile water and drying off. The surface decontamination process should not expose the contents to antimicrobial factors such as elevated temperatures. Equipment to be passed into the workstation should be sterile. Rinse fluids should be sterile, for single use and packed in tamper-evident containers where external surfaces are sterile or of known good microbiological quality. They should be released for use by a formal procedure that checks the sterilization cycle, formulation and expiry date. More than one positive sterility test result has been traced back to inadequately controlled rinse fluids.

Sanitization fluids may be another source of contamination or inadequate performance. If alcohol is used it should be ethyl or isopropyl alcohol, assured to be free of viable micro-organisms, including fungal and bacterial spores. The alcohol should be diluted to about 70% with sterile water and put into sterile spray bottles. Other sanitization agents should be similarly free of micro-organisms, and suitable in use expiry dates should be set.

Transfer of transient items to the workstation is a critical operation since micro-organisms may gain access to the otherwise microbially controlled zone. Surfaces exposed to contact or airborne contamination need to be treated. Ideally, items from autoclaves should be bagged with packaging being removed immediately prior to the contents being introduced into the workstation. After autoclaving, large items like vacuum manifolds suck air through the bagging material on cooling. The bagging material is often a poor microbial filter, so any air sucked in should be of good initial quality; ideally, such items should be cooled in laminar airflow. Items which cannot be autoclaved, e.g. samples to be tested or purchased pre-sterilized filter assemblies, should be transferred to the workstation by the safest route possible. Where purchased assemblies have been sterilized in an over-pouch, it may be sufficient to remove the over-pouch as the assembly is passed into the workstation. On the other hand, if the pouch provides the sterility barrier to the assembly, it will have to be removed inside the workstation and just before use.

If sterility cannot be assured on outer surfaces, surface sanitization will be required prior to entry to the workstation. This is probably the weakest part of the process. Sanitization with agents such as 70% alcohol is far from perfect. If the alcohol makes contact with micro-organisms it will kill a high proportion of most types, but if contact with alcohol is impeded by dirt it will not be effective. If the organism is a bacterial endospore it will not be killed, even if contacted effectively. Thus, the surface should first be cleaned and then covered in sterile 70% alcohol using a spray, dunking baths or alcohol-soaked wipe. The surface should be systematically wiped once in one direction with a sterile swab, thereby physically removing any remaining viable cells. This is a slow deliberate wiping operation, usually starting at the critical point, for example the septum of a vial and working systematically to the base.

Gloves used for handling during this transfer stage should initially be sterile and frequently sanitized thereafter. Glove integrity should be visually checked and if alcohol leakage is noted into the inside of the glove it should be changed. If the operator moves away from the room new gloves should be put on during re-entry.

### 8.2.8 *Workstation*

Sterile gloves should be worn inside the workstations which, together with its resident equipment should be thoroughly cleaned and free of visible deposits. The internal surfaces should be sanitized using sterile swabs, e.g. 15 cm × 15 cm, soaked with sanitization fluid and wiped using a serpentine motion working away from the air filter (a horizontal flow laminar airflow hood requires at least four such swabs). Any resident equipment in the workstation should be thoroughly swabbed using a fresh swab for each item. If the sanitization fluid is not volatile, the surfaces should be rinsed with sterile water to remove fluid residues of the sanitization agent and then dried off. During cleaning the placement of any removable sides in the workstation should be checked. If improperly located, contaminated air from the room could be sucked into the workstation with an increased risk of contamination where aseptic manipulations take place. Similarly, holes in the side panels for access of gas or vacuum tubes may permit leakage of room air if improperly sealed.

The cleaning and sanitization programme of the workstation should include a daily clean with detergent and rinse, avoiding wetting the high efficiency particulate absorption (HEPA) filter. Sanitization using volatile agents such as 70% alcohol should be carried out before each test. A weekly thorough detergent clean should be followed by sanitization using a sporicidal agent such as 500 parts per million free chlorine sodium hypochlorite solution, glutaraldehyde or other proprietary sporicidal agents (further information on sanitizers and disinfection appears in Chapter 12). After application and the appropriate contact time, the sporicidal agents should be thoroughly rinsed off and the surfaces dried. If novel sporicidal agents are used it is advisable to confirm manufacturer's claims under the conditions of use.

With the use of isolators another level of assurance can be gained if gassing can be employed. Provided that the isolator is positively pressurized with HEPA-filtered air and is leak free, the ideal environment for aseptic manipulation can be created. A significant cause of false positives is inadequate preparation of the workstation and insufficient surface sanitization of items taken into it. In such isolators sporicidal gas, e.g. peracetic acid or hydrogen peroxide, is usually introduced through the inlet air filters. The gassing cycle should be validated and this is usually carried out using biological indicators. A six-log reduction of the indicator organism provides a good assurance of lethality for the types and numbers of micro-organisms normally encountered in a well-cleaned isolator and on the outside of samples and equipment. As gas may not penetrate between touching surfaces, some internal rearrangement or support on point contacts may be required. The possibility that the gas has penetrated samples under test or containers of culture media, with a consequent risk that the test may be invalidated, should be checked.

### 8.2.9 *Aseptic manoeuvres*

The sterility test may involve placing broth into containers either to test the inner surface of the device under test or to fill the incubation container that will hold the test filter. Alternatively, there may be a flushing process applied to surfaces of a device followed by filtration or other assay of the rinsings. This flushing technique is a poor way to carry out a sterility test because of its inherent inadequacies in removing and transferring all micro-organisms present to the culture media. A closed technique is preferable for transferring fluids, i.e. connections should be made between bags, bottles and containers by tubes fitting into or onto properly designed ports or by spiking through rubber membranes. Transfer from ampoules to sterility test sets cannot be done completely using closed techniques and there may be a trade-off in risks between the assembly of a syringe and needle and simply breaking open the ampoule and pouring the contents into a filter holder or broth container. As soon as pouring techniques or ampoule handling becomes necessary, the use of down-flow work stations adds to the risk of false positives, since these techniques require hands to be positioned such that they intervene between airflow direction and critical open surfaces.

Simple readily cleaned and sterilized devices, e.g. wire cradles, supports, stands, will leave hands as free as possible during aseptic techniques and avoid small items falling over. For tops and closures, a well-designed carrier at the proper height, ensuring sterile surfaces are not compromised, may be useful.

The basic principles of aseptic techniques are as follows:

1  Critical, i.e. sterile surfaces, should be physically protected whenever possible and only exposed for the minimum time necessary to carry out the manipulation.

2  Operators should move at a pace that reduces air turbulence and does not result in mechanical shocks, but which nevertheless minimizes the time that critical surfaces are exposed.

3  Critical surfaces should not touch non-sterile surfaces (note that surfaces sanitized with 70% alcohol are not expected to be sterile).

4  When a critical surface is exposed, the laminar airflow should have clear access to exclude micro-organisms; impeding the airflow by the direct intervention of hands or equipment must be avoided, or if this is impossible the intervening object should have sterile surfaces. Turbulence caused by objects near the direct line of laminar flow can also lead to contamination or cause back-flow, drawing contamination from downstream to return against the mass flow. Similarly, working too near the front of a horizontal flow hood (within 30 cm or so) or too near the base of a down-flow hood (within 15 cm) means that laminar flow is less likely.

The practical problem is that, for example, taking a sterile hypodermic needle from its paper package involves gripping small fins to split the pack open; this immediately exposes the critical surface (the open Luer connector of the needle) to turbulent air from the fingers only a few millimetres away. Also, removing some closures from bag ports, spikes, etc. brings fingers (albeit gloved) into close proximity to critical sites. Careful choice of easy-to-use equipment and using optimized aseptic techniques can repay the extra cost in reduced false-positive rates.

### 8.2.10  *Labelling*

If a fluid or item is transferred from a labelled container to an unlabelled container in continuous sight of, and under the control of, a trained operator, it could be considered to retain its identity by virtue of the operator's memory and continuous attention for a short period of time, e.g. seconds; after this short period the container must be labelled or the fluid or item loses its identity. It is often tempting to try and recover a loss of identity situation by a process of elimination, but the job of the sterility tester is to move in control from one known situation to the next known situation. Guessing does not play a part in good laboratory practice, however good the circumstantial evidence. Written procedures should recognize that loss of identity will sometimes take place by accident and have (non-blaming) recovery methods. The sterility tester should be trained that the company would prefer to discard a valuable batch rather than release one unsterile unit of product.

All containers should be labelled before work begins. Labels and work documents can be a potential source of micro-organisms, including bacterial endospores. Ideally, labels present in the workstation, or to be otherwise handled, should be sterilized by irradiation or autoclaving. Similarly, pens should be dedicated to particular uses.

### 8.2.11  *Filter holders and filtration*

Sterile ready-to-use sterility test sets may be applicable, otherwise 0.45 μm or 0.2 μm rated filters can be sterilized already assembled into filter holders. The filter should be easily detached from the filter holder after sterilization *in situ*. Transfer of pre-sterilized filters to separate filter holders increases the risk of contamination and false positives. Open-topped filter holders can be accidentally contaminated, so their use should be a last resort.

Autoclaving filters previously assembled into filter holders can result in burst filters. The filters should be assembled bone-dry with as little rotational stress as possible. The assembly needs to be covered or bagged in steam-permeable material. The validated autoclave should have a porous load cycle with gentle evacuation stages, that do not over-stress the filter, in order to remove air prior to admission of steam and for drying phases at the end. Return of air to the autoclave at cycle end should be via a micro-organism-retentive filter which is properly maintained. Gravity displacement autoclaves are insufficiently reliable for this task. As previously indicated, after autoclaving the bagged filter assembly should be allowed to cool in HEPA-filtered laminar airflow to reduce the risk of contaminated air being sucked into the assembly. The management of autoclaved and unautoclaved bagged assemblies needs to be rigorously controlled in order to avoid an unautoclaved assembly being used for testing. It should not be assumed that negative controls will enable the detection of such errors.

Sufficient assemblies should be available to allow their recycling in a well-controlled way; rushed cleaning and autoclaving leads to problems. After use, the washing process should provide for adequate cleaning and proper rinsing to remove traces of antimicrobial detergent. The components should be rapidly dried to prevent micro-biological build-up in regions that do not drain effectively. These difficulties in preparing filters for sterility testing increase the process risk. Laboratory preparation may be suitable for less critical bioburden testing, but ready-to-use sterile sets are safer for sterility testing.

Fluid preparation and the process of filtration should follow pharmacopoeial recommendations. Vacuum filtration is most often employed, although the use of positive-pressure filtration using sterile filtered air or gas is possible. As vacuum pumps can 'hiccup' back non-sterile material and as filtrate fluids need to be properly managed, a reservoir container is required in the vacuum supply line outside the workstation or outside the room. A routine of autoclaving, replacement, washing and sanitization of the vacuum lines is necessary to remove potential sources of micro-organisms from the critical work zones.

Filter assemblies may be connected to or placed onto a manifold so that several filtrations can be carried out simultaneously. The cleaning and autoclaving of such manifolds is another potential source of difficulty which can leave micro-organisms contaminating valve assemblies within a few centimetres of the filter. Valves may need dissembling to be properly cleaned and sterilized.

## 8.2.12  Media

After filtration, the filter is incubated in contact with the broths recommended in the pharmacopoeia. If broths are purchased sterile and ready to use, the user should ensure that they are of suitable quality by auditing the manufacturer's quality system. Fertility tests and negative controls should only be regarded as crude quality indicators. Some questions that may arise during audit of a media manufacturer are listed in Chapter 14.

Given a satisfactory audit, the user should have a system of written procedures for checking media on receipt, scrutinizing expiry dates, batch numbers, release documents, fertility tests, etc. The user should also decide what positive and negative controls will be applied to the broth batches and sterility test runs with a written rationale. Where media are prepared from commercially available ingredients, the same controls expected of manufacturers should be applied. For media such as fluid thioglycollate, the significance of the colour change of the resazurin redox dye should be fully understood and appropriate limits set.

## 8.2.13  Incubation

Temperatures recommended in the pharmacopoeia should prevail at all points in the incubator. Temperature mapping and proper maintenance and control of equipment is a prerequisite. The temperatures should be monitored, recorded and checked throughout incubation from representative positions using calibrated equipment. Cultures should be inspected daily *in situ* to enable early detection of obvious growth and to instigate follow-up action. If incubator temperatures exceed their limits, written procedures should clearly specify what action is to be taken.

Low temperatures can be compensated for by prolonged incubation for equivalent periods unless this reduces the ability of the test to detect indigenous flora or renders the broth less likely to support growth. Low temperatures may induce interruptions of growth and lag phases before growth resumes. High temperatures are less easily dealt with, but short periods over the set value may not result in a significant rise in the temperature of the broth. The microbiologist should construct a proper rationale for actions specified in the written procedure based on local methods, history and flora. However well maintained, incubators on rare occasions may fail. Temperature fluctuation outside limits, however compensated for, always raises questions about the validity of the sterility test. Such excursions are likely to be viewed with greater suspicion in an aseptic rather than a terminally sterilized product.

The period of incubation is a vexed subject. On one hand, the extension of the period imposes penalties on those that run well-controlled terminal sterilization processes with vast safety margins. On the other hand, prolonged incubation increases the chance of detection of faulty product and processes from weak aseptic operations. Pharmacopoeial recommendations have to be respected or other approaches justified.

### 8.2.14  *Examination*

Visible evidence of growth may be a general cloudiness, flocculation particles, surface pellicle or scum, fibrous pellets or colonies seen by a person whose eyesight is
verified annually. In order to detect growth the incubated broth should be gently
shaken and compared with a negative control or original broth. The final comparative viewing should be a formal operation with defined conditions of illumination,
e.g. a viewing booth or trailing lamp used in a specified way.

### 8.2.15  *Interpretation*

If cloudiness or particles are present, the container integrity should be examined
carefully for any obvious signs of weakness, e.g. cracks or displaced injection sites,
that may justify a declaration that growth could have resulted from contamination
after testing. The container may then be subject to pressure testing to confirm
suspicions or reveal weaknesses. Careful aseptic sampling of the fluid and microscopic examination in experienced hands may throw an entirely different light on the
source of cloudiness or particles, justifying a declaration of an invalid test due to
obvious chemical precipitation. The cause of such invalidation and its subsequent
prevention should be investigated retrospectively to justify re-testing. The suspect
broth should be subcultured, plated onto relevant solid media, and incubated both
aerobically and anaerobically. Where chemical precipitation is suspected, absence of
growth is supportive, but when microscopic examination reveals the presence of
micro-organisms, absence of growth in the subculture cannot be interpreted as
lack of evidence of microbial contamination.

Growth on solid media from either suspect chemical precipitation or suspect
microbial growth is clear evidence of microbial growth in the broth. Some apparently
crystalline or amorphous precipitates are due to microbial action.

At this stage of the investigation a declaration that the results of culture on solid
media are due to accidental laboratory contamination risks losing all credibility from
the operations, and is unlikely to be accepted unless exceptional circumstances prevail.
For this reason, the follow-up of a putative positive sterility test should be conducted
by a qualified microbiologist with current bench experience. If faults in methods, as
demonstrated by contaminated rinsing fluids or growth in negative controls or lack of
integrity of culture container, fail to provide an explanation, then the possibility of a
genuine sterility test failure has to be considered.

It is a reasonable, though conservative, view that if the sample tested was from an
aseptic process, the justification for re-testing has to be particularly strong. Collecting
information that supports a case that the test was invalid based on environmental
contaminants found in a low-grade or poorly controlled sterility testing laboratory is
not very convincing. If the organization or company elects to carry out sterility testing
in poor facilities where method contamination is a real possibility, then the consequences must be accepted.

The fact that the sterility test contaminants have not been found in the aseptic filling
process or environment, but have been found in the sterility test facility, is considered a
weak argument. Such a case for discounting the original test and re-testing would have
to involve exceptional types of micro-organisms to gain credibility. In addition, if the

sterility test was carried out in a properly run isolator there are few grounds for re-testing samples from an aseptic process.

In summary, if the sterility test of samples from an aseptic process shows growth there are very few good reasons to consider the first test invalid and to justify re-testing. Positive results from tests on products from terminal sterilization processes may allow different arguments and considerations to prevail. If the sterilization process is a conventional one, e.g. dry heat, autoclaving, irradiation or ethylene oxide gassing using processes containing huge safety margins, then the cause of the positive growth is narrowed down. Possible causes are:

- product not exposed to the sterilization process;
- sterilization process was not delivered;
- loss of product integrity after sterilization;
- the survivor is an extraordinarily resistant organism; and
- the sterility test method introduced the contaminant.

For materials that can be resterilized, i.e. are not damaged by the process and inherently cannot support microbial growth during the lag time between completion of production and final effective sterilization process, loss of product may not be at risk, but the cause of the positive test still needs to be found. When materials cannot be resterilized the batch is lost if the first four reasons apply; however, if these reasons can be shown not to be the cause, the possibility that the sterility test was faulty could be considered. Proof of a faulty test may arise from the general considerations of failed negative controls, lack of integrity of the culture container and the other reasons outlined above. A history of low level failures, i.e. less than 0.1% of cultures incubated, provides circumstantial supporting evidence; however, a high level of such failures or none at all does not support a decision to re-test. Thus, frequent positive results indicate that the facility possibly produces non-sterile product, or that there may be a lack of sufficiently trained or experienced staff to assure the sterility of products. If no positives are normally seen, then a positive test is a very significant event and its cause should be determined. If the rationale is sound, then re-testing in compliance with the pharmacopoeia is appropriate for terminally sterilized product.

One of the main concerns in evaluating sterility test positives is to err on the side of caution from the final user's point of view. There is a paradoxical consequence that, if a batch is discarded due to a sterility test deemed to have failed, the process used up to that point is effectively invalidated. Unless a discrete cause is found that can be permanently corrected, the wisdom of the continued operation of the process and use of existing product should be questioned.

### 8.2.16 *Documentation*

Documentation of the sterility test should be considered as the means by which the history of the process can be reconstructed. Unexpected results or deviation from the norm should be treated with caution and documented carefully; they may well justify a re-test, but if an incident occurs which could explain and invalidate a positive result, there is an argument to abort the test process and repeat with fresh samples.

The notion that if a test passes it will be accepted as valid, but, if it fails, an invalid test will be declared does not foster confidence.

### 8.2.17  Summary

The optimum approach to sterility testing can be argued to include the following considerations:

1  People require inherent skills of manual dexterity, visual acuity, and ability to sustain concentration. They should be properly trained in aseptic manoeuvres and continuous working sessions should be limited to 2 h. They should be supervised by properly qualified supportive management who provide proper resources and work to high levels of integrity.

2  The ideal facility is a good quality, positive-pressure isolator. A horizontal laminar flow hood in a Grade B clean room can also achieve good results, but a downflow hood should only be used after careful consideration of the disadvantages.

3  An effective sanitization method is peracetic acid or hydrogen peroxide gassing in a positive-pressure isolator. A less effective alternative is 70% alcohol, and it must be properly applied to obtain the best results. Volatile quick-acting sporicidal sprays are now available, but have yet to be validated at the time of writing.

4  Aseptic manipulation should employ closed systems whenever possible. Open techniques, e.g. open top filter holders, pouring from bottles, etc. are more prone to false positives.

## 8.3  Bioburden testing

### 8.3.1  Scope

Bioburden testing is used to estimate the number of micro-organisms in or on the following: product prior to sterilization (pre-sterilization count); solid components and process aids; liquid components and process aids; chemical raw materials; and gaseous components.

### 8.3.2  Methods and principles

Good laboratory practice and general laboratory principles apply as with sterility testing. Details of methods for examining medical devices are described in EN 1174 (1996, 1997) (also discussed in Chapter 3), and may be suitably modified for pharmaceuticals. The following should be considered:

1  Does the method of sample preparation result in transfer of antimicrobial substances? This can be tested by seeding relevant media and rinse fluids with known numbers of standard fertility test micro-organisms or species known to be sensitive to substances likely to be present. After an appropriate contact time the number of survivors can be determined. If antimicrobial substances are present, they should be neutralized, removed, e.g. by filtration, or diluted out.

2  What is the relationship between the number of micro-organisms present and the number recovered by the test method? The main method involves seeding the material with known levels of indicator organisms (e.g. drying bacterial spores onto a device) and determining the number recovered by the chosen recovery methods. Another method, more specifically for convenient surfaces, utilizes the recovery by overlaying the surface with agar medium to determine the number of survivors.

3  Do the culture techniques achieve the desired objectives? If a non-specific aerobic count is required, are the media, incubation time and temperature ideal for the likely flora? Those providing the best recovery conditions are selected. Culture for specific genera can follow current practices described in text books and media manufacturers' literature, and if necessary be checked by control organisms.

Pre-sterilization counts carried out in support of specific sterilization processes can involve selective sample treatments. For example, the presence of heat-resistant spores is important information for autoclave or dry heat sterilization processes. For the manufacture of liquid preparations sterilized by autoclaving, the screening of chemical raw materials, process streams and product prior to sterilization can readily be carried out by raising the solution temperature to 80°C for 20 min and plating or filtering as normal. The survivors are putative spore-formers and further tests of heat resistance can be carried out on sporing subcultures, e.g. an exposure of 100 min at a 100°C applied to 10 000 bacterial endospores selects out types with a heat resistance higher than normal. The heat resistance of the original spore in a pre-sterilization count sample can never be known because the spores tested are subcultures. The best that can be done is to characterize the culture and hopefully show it is not from a high heat-resistant species. Another factor to take into account is the possibility that the heat resistance of the spores varies depending on the nature of the suspending fluid, so tests of resistance in the finished product solution may be appropriate.

### 8.3.3  *Sampling*

The sampling of chemical raw materials should not result in contamination of either the bulk material or sample. The use of filtered airflow booths and sterile or sanitized equipment is advised. Sampling of process streams ideally avoids the use of special sample valves or any other equipment that may add contamination to the sample or process stream. Sampling from natural outlets of process streams is recommended. Failing this, the sanitization and flushing of sample points is necessary. From experience, as much as 10% of 'out-of-limits' results are eventually explained by faulty sampling or old contaminated valves. In such circumstances the risk that these valves present to the process stream should be considered. Similar arguments apply to waters, but fortunately, natural usage points are often suitable for sampling. Lubricants, solvents and other fluids used in the process are often from small-scale containers and present few sampling problems. Containers for collecting samples should be sterile and properly labelled. When sample containers are prepared with neutralizing agents (e.g. sodium thiosulphate for chlorinated waters), these too should be properly controlled, i.e. batch number, expiry, documented process, release and stock control.

Samples of solid componentry taken during the production process should provide some representation of both worst case and typical conditions.

Gaseous streams, such as compressed air or gas cylinders, rarely cause microbiological problems provided that the systems are properly designed and maintained to avoid water ingress or build-up, and the correct filters are in place. Occasional sampling by bubbling known volumes through broth or by capture on special gelatin filters is worthwhile.

Testing pre-sterilization samples plays a major part in most sterility assurance systems as the results show process stability and the continued justification of the sterilization process originally validated and routinely used. Products that are prone to microbiological contamination or spoilage, including aqueous products with or without preservation systems, have to be carefully managed. The time limit from the mixing stage to the start of sterilization should be as short as possible, and certainly no longer than 24 h, to reduce the risk of microbial growth and challenge to the sterilization process. The reduction in the opportunity for growth of bacteria also reduces the risk of development of endotoxin.

In order to determine the count just prior to sterilization, a sample requires testing at the time sterilization starts. In practice this may be difficult, resulting in a compromise to refrigerate the sample at this time. This is a poor compromise because chilling may be slow and growth may continue; also after storage at low temperature the bioburden may not regrow out when tested. In addition, micro-organisms may become attached to the surfaces of the container under these static conditions and be difficult to resuspend for sampling. Storage time in the refrigerator should be limited to the minimum necessary. A peculiar feature of microbiologically unstable pre-sterilization samples is that resampling is not an option; retention of samples in the refrigerator reduces their relevance as time elapses. As a consequence if the refrigerator develops a fault or samples are mislabelled or lost, the situation is difficult to recover. Written procedures should cover this possibility, and as long as process stability can be demonstrated, this may provide justification to waive the need for pre-sterilization counts prior to batch release in exceptional circumstances.

Accounting for pre-sterilization samples generally is of critical importance. The number withdrawn from the batch and relabelled for testing, the numbers placed in the refrigerator or collection point, withdrawn from the collection point, tested and the surplus destroyed should be recorded diligently for each batch. In addition, the company should have a demonstrable policy that samples withdrawn from the batch at any stage are never returned. These precautions should reduce the risk that unsterilized samples find their way back into saleable stock. The pre-sterilization samples should be taken from worst case parts of the process. Attempts to define statistically based methods of determining the numbers of samples needed may fail because the distribution of counts rarely follows a predictable mathematical distribution. In spite of this the sampling regime should be well thought out with a fully documented rationale.

### 8.3.4  *Limits*

The setting of limits, interpretation of results and use of data generally is the harvest of the combined preceding effort. The objectives of bioburden testing include the following: to explore process stability; to meet limits imposed by outside bodies; and to detect excursions that may affect the safety of the product. Each process will produce a spread of microbiological counts when it is considered to be in control. If this spread can be fitted to a recognized mathematical distribution, statistical techniques can be used to set limits, e.g. at the 99% or 95% level of probability.

Often a mathematical distribution model cannot be fitted to data and an empirical approach is used. In this method the data for the preceding year are analysed and the count above which 5% or 1% of results occurred is determined. This count is assessed to see if it would provide a useful limit to enable lack of process stability to be judged. If the potential limit is greater than the limit set in the preceding year it would be discarded, otherwise a continuing deterioration in batch quality could be accepted. If the potential limit is smaller than the preceding year, the wisdom of setting tight limits from an atypical year needs to be considered. If implemented changes have reduced the spread of results, then using a tighter limit is reasonable.

The relationships between safety limits, externally imposed limits and in-house process control limits should be evaluated. A certain amount of leeway is advisable, otherwise the microbiology section may over-react to fairly irrelevant fluctuations in counts. Limits imposed by customers, regulatory authorities, head office, etc. may be successfully questioned based on detailed knowledge of the performance of the process and its true safety limits, but often must be observed.

Should in-house process control limits exceed imposed limits, a process change may be required. Safety limits are usually far in excess of in-house control limits. For example, in steam sterilization counts of a thousand heat-resistant bacterial endo-spores per unit prior to autoclaving may be necessary to pose a risk of non-sterility.

The possibility of emergence of particularly resistant types is always present; with ethylene oxide gassing, *Pyronema domesticum* from heavily infected cotton was found to survive standard cycles in recent years. It is incumbent upon the microbiologist to ascertain the true safety limits of the product and properly relate them to the in-house process control and imposed limits.

### 8.3.5 *Interpretation of results*

Having set limits, written procedures should clearly describe actions to be taken when they are exceeded to avoid any uncertainty. Such actions may involve issuing 'out-of-limits documentation' demanding a properly documented follow-up and corrective action. Sometimes the only course of action is to discard the batch of product. Here, written procedures, i.e. a contract between the production and quality assurance groups provides clear guidance as to how to proceed in defined and often stressful circumstances.

The nature of microbiological control is that if it fails, the counts are usually obviously high and do not need sophisticated cusum analysis or even simple graphs to show that the process has become out of control. The problem often is that it is a short-lived event. Incidentally, it should be a matter of honour for the microbiologist to come to a reasonable theory to account for failures, even if clear proof is elusive.

The normal method of detection of a process that is out of control is by the experienced operator reading the plates and scanning adjacent data by eye. This can be supplemented by graphical representation of data over time to show up long-term slow deterioration or to detect seasonal trends which can add to the understanding of the process. Rolling averages, when 12 months' or 3 months' data are averaged and plotted and then rearranged for the next months' data when a new 12-month or 3-month average is plotted, can smooth out short-term perturbations and reveal underlying long-term trends.

### 8.3.6  *Characterization of flora*

Changes in microbial flora may be significant. General surveillance and character-ization of flora in 'out-of-limits' situations is useful. Experienced operators recognize colonies as normal or abnormal. In addition to colony appearance, microscopic examinations complemented by staining may enable characterization as follows:

1   Moulds can be categorized by experienced staff into several putative genera based on colony characteristics and low-power microscopic examination.

2   Yeasts can be distinguished on the basis of colony morphology and microscopic examination.

3   Bacteria may be classified as Gram-positive, Gram-negative, rods, cocci, clusters and spores allowing initial categorization.

4   Algae, protozoa, crystals, amorphous material, starch granules, plant parts, blood cells can all be readily differentiated by microscopic examination.

Here, the experienced industrial microbiologist can evaluate the significance of a change in the flora and perhaps prevent closure of an operation by taking pre-emptive action. Infestations of clean rooms with fungal spores or bacterial spores continue to take place and the slightest sign of such an event should be treated seriously.

### 8.3.7  *Recording and reporting data*

One advantage of computer-stored microbiological data is that sophisticated trend analysis of the selected series of data, if needed, is available. A disadvantage is that scanning recent history en masse for patterns, as opposed to selected series, is often not an option. Summaries of bioburden data are often produced and provide a useful accessible historical record to help evaluate long-term changes; however, for non-microbiologists the summary has its shortcomings. A summary of 'out-of-limit' situations or significant excursions and events may be of more use.

   Results from bioburden determinations should be collated and brought to the attention of senior management as: significant effort has gone into data collection; bioburden data are a critical part of sterility assurance; and the data are another indicator of factory performance which may tie in with other changes that are not known to the microbiologists.

## 8.4  Biological indicators

### 8.4.1  *Introduction*

Biological indicators (BIs) are devices containing a known number of micro-organisms of a known strain and resistance to an inactivation process. Reference should be made to the *European Pharmacopoeia* and EN 866 (1997) for guidance relevant to pharmaceuticals and medical devices respectively. BIs are used to assess the lethality or degree of microbiological inactivation delivered by a process. As part of the estimate of the sterility assurance of a system, the relationship between the resistance of the BI and the likely resistance of the naturally occurring micro-organisms should be known or estimated. The subject of use of BIs as a bioburden model cannot be described in detail in this chapter, but further information is available elsewhere (Denyer and Hodges, 1998; Baird, 1999). The effort to disentangle concepts of sterility assurance from kills of standard BIs and relate these to pharmacopoeial sterilization cycles and resistance of bioburdens is a prerequisite for microbiologists involved with sterility assurance systems.

### 8.4.2 Applications

BIs are used when delivery of a microbiological inactivation process cannot be physically measured. They may be used in validation or as routine process monitors. Such situations include: ethylene oxide sterilization; low-temperature steam and formaldehyde; isolator gassing systems; steam, when penetration of steam is not physically assured (autoclaving or sterilization in place); and irradiation to aid the study of micro-environments.

### 8.4.3 Sampling

Sampling should detect weaknesses in the inactivation process itself. Sometimes protocols are set up to generate data showing that the system is satisfactory, rather than providing a rigorous challenge. Numbers of BIs are no substitute for intelligent placement based on knowledge or theories about weak positions; nevertheless, the whole sterilizing chamber still needs surveying in case of unexpected problems.

Issues which may be of interest are:

- In ethylene oxide and other gassing systems does the placement of BIs explore the effects of rapid de-gassing at cycle end, as well as adequate build-up of gas concentration?

- In all systems where evacuation is a feature, does the test explore the possibility that residual air in partially contained systems, such as tubes or vented bags, is chased back to a zone where moisture permeation, steam penetration, gas penetration, etc. is inhibited?

- Are micro-environments in loads for irradiation adequately surveyed by large dosimeters? Should BIs or smaller dosimeters explore shielded areas or areas where microbial survival may be enhanced?

- In steam sterilization, how is air removed and replaced with steam if permeation of steam through plastics into the product is involved?

- In dry heat sterilization what is the insulating effect of component materials?

- In gassing isolators does gas penetrate between surfaces in contact or into zones such as dead-end tubes?

- BIs can be used to explore areas considered at particular risk, such as the interface between the vial and its rubber stopper; the dry areas at the top of ampoules; the contact points of the syringe plunger with the barrel; needle lumens; plastic-to-plastic interface surfaces in devices or plastic-encapsulated air pockets in bag ports.

### 8.4.4  *Handling*

Biological indicators should be stored in compliance with the manufacturer's instructions, including a specified relative humidity, and discarded on expiry. The correct BI should be selected for its intended purpose. Those held within impervious containers should not be used if penetration of moisture or gas during sterilization is part of the mechanism of inactivating micro-organisms in the product.

BIs often have protective packaging which reduces the risk of spore liberation of the BI organism. If this cover is removed, not only is there a risk of spores contaminating the environment, but the certified resistance of the BI may also be changed.

For validating autoclaving of cartridge filters for aseptic processing there is a case for precipitating BI spores onto and into the filter material. In these circumstances the maintenance of the certified decimal reduction time (D value) requires the manufacturer's guidance to be followed carefully.

Suspending or locating BIs in the desired space often involves ingenuity, including the use of fishing line, wire, tape, or for central placement, a special test section sealed into long tubes. All these methods carry a risk of changing the way in which the inactivation agent penetrates the test site, so proof that a worst case location is not changed into a best case location by the method of placement is advised. A basic requirement is that all BI units put into a system should be recovered; a resistant microbial preparation lost in a production environment is an unnecessary risk.

Removal of BIs from the load after treatment should be carried out in an appropriate way. Gassed systems, e.g. ethylene oxide, hydrogen peroxide and peracetic acid, may continue to leach out gas absorbed into plastics, paper, etc. long after the controlled sterilization process has finished. The residual lethality can be equal to the original cycle, but is usually uncontrolled and cannot be used to assure sterility. For this reason BIs should be recovered as soon as possible after the end of cycle (within minutes if possible) and subjected to a process that eliminates residual lethality, e.g. refrigeration, immersion in sufficient diluent, or neutralization, all of which need validating.

In the event of an aborted cycle, great care needs to be taken with BIs already placed in their test positions. Preliminary stages of gas and autoclave cycles can change the BI sensitivity, so BIs may need replacing in the event of abort situations.

### 8.4.5  *Culture evaluation and interpretation of survivors*

Microbiology laboratories handling BIs run the risk that highly resistant micro-organisms may escape to other parts of the laboratory, e.g. bioburden and environmental testing sections. When handling positive controls or BIs where survivors may be expected, separate laboratories or containment cabinets are used.

The recovery system should support the growth of the BI organism after it has been exposed to a sublethal process, e.g. a very short cycle. If this control does not show growth, the test is invalidated. Two main approaches to culturing survivors are currently employed. The first, a 'drop' test, involves dropping the BI after taking off any envelopes into the recommended recovery broth and incubating for the development of turbidity. The second method involves subjecting the unwrapped BI to disintegration in the case of paper carriers, or shaking/sonication for removal from carriers that will not break down. The resultant suspension of organisms can then be plated using conventional techniques to count survivors.

Drop tests only provide evidence of kill or survival. Thus, if ten BIs have been distributed in the load and exposed to the process and all show no growth, clear conclusions can be drawn. However, a fraction negative result, e.g. one out of ten positive, cannot be advantageously interpreted by recourse to statistical considerations, e.g. most probable number calculations. The underlying principle for the proper application of these statistical methods is that ten BIs have been exposed to exactly the same process of inactivation. This is not a reasonable assumption when, for example, an ethylene oxide sterilizer or autoclave has been seeded with BIs in different positions specifically to find out if homogeneous lethality prevails. The one positive BI may be in an area that just does not experience the same concentration of gas or penetration of steam, and this is why the test was first set up. However, if three or four BIs were put in each of the ten positions, other considerations can be used because in each position homogeneity may reasonably be expected. The best approach is enumeration of survivors, but it is more demanding on laboratory resources. In this method the biological indicator that would just give a positive result if subjected to a drop test, may be revealed as containing 100 survivors and a truly weak location in the sterilizer is thus revealed.

There is another aspect which should be noted; experience shows that in situations where a drop test in broth shows a positive result, enumeration often shows a zero count. From this it may be speculated that a single damaged survivor may recover and grow to produce turbidity in broth, but would not produce a visible colony on the solid medium used for enumeration. It is possible that incubation of a filter on a broth-soaked pad may enable enumeration at low survivor levels that would not be detected on agar medium. The significance of this phenomenon in sterilization processes which incorporate generous safety margins is not likely to be great.

Enumeration enables a graph of survivors versus extent of process to be plotted, and thus the course of inactivation can be defined. By a process of extrapolation and confirmatory testing, the necessary minimum extent of the process of inactivation can be deduced. Cycles can then be defined by time or gas concentration plus time, or irradiation dose, etc. after a suitable safety margin has been added.

### 8.4.6 *Miscellaneous considerations*

A presumption made above is that the survivors are the BI organisms. Should another organism grow, this poses a question whether or not this was a random contaminant of the recovery stage, or was it there originally and so resistant that it survived the process? In these situations the veracity of the data is in doubt. Could this contaminant have masked a genuine BI survivor, for example? In order to retain credibility, there should be no likelihood of chance contaminants.

Methods for handling routine process BIs should follow published guidelines. The inclusion of positive and negative controls does at times yield useful information: a situation where the positive control takes longer than normal to outgrow, or one tube is negative and the others are positive, should be regarded as unacceptable. Traces of detergent left in tubes can prevent outgrowth and invalidate the test.


## 8.5   Environmental monitoring

### 8.5.1   *General*

Clean rooms, laminar flow workstations and isolators with their associated procedures and disciplines are all methods used to reduce the risk of micro-organisms being present at the point where critical operations take place. The success of these arrangements is dependent not only on the design, but also on continued correct functioning and on adherence to procedures. For this reason the assurance of continued delivery of safe conditions for critical operations rests heavily on planned maintenance, physical checks of correct performance, calibrations and supervision of well-trained staff. The staff of microbiology laboratories can contribute to training and auditing.

In addition to these activities a considerable and possibly disproportionate emphasis is placed on environmental monitoring. Sometimes this has led to such reliance on environmental monitoring data that poorly designed, maintained and run systems are claimed to be satisfactory based on good environmental monitoring results.

The methods usually used for microbiological monitoring are settle plates, active air sampling, contact plates and swabs for surfaces and garments, plates for finger dabs, and although not directly microbiological, particle counting for airborne particles.

These methods only provide an approximate guide to the level of micro-organisms present from a small sample of space, surface and time. As such, they make a contribution to the knowledge of the true microbiological state of a clean room, laminar air flow cabinet or isolator. It follows that the best use should be made of such tests without being too pedantic about their precision or accuracy.

The history of 'in-limit' results from properly designed and executed environmental tests provides confidence that the primary measures of control (maintenance of properly designed systems, training and supervision of staff) are producing a stable situation. More information, however, is provided when 'out-of-limit' results are found suggesting lack of process stability. Causes should be investigated with a high priority and concluded with corrective actions.

Isolated counts just above limits (see section 8.4.5) may never have reasonable causes assigned, but where this occurs more frequently or gross failures occur, a detailed investigation followed by corrective action is expected. Even when isolated counts just above limits are found, the characterization of the micro-organism may provide a reasonable clue to the possible source.


### 8.5.2   *Sampling*

Environmental sampling should not be regarded merely as a ritual gesture of monitoring to confirm that the system is in control. Complimenting staff who find problems as well as recognizing staff who solve problems helps to create the proper psychological approach to building assurance of environmental microbiological control. Sampling plans should include some optional *ad hoc* undefined sites as well as routine challenging sites such as the wheels and axles of trolleys and mobile tanks. Other challenging sites include the under-sides of tables and stools, door hinges, switches, tops of doors, public address loudspeakers, faults in laminates and vinyl surfaces and stains that might originate from leaks or drips. The people who routinely clean the area know the difficult places, as do the people who run the machines, so it is sensible to seek their advice before defining sites for sampling.

Environmental sampling considerations are discussed further in Chapter 3.

### 8.5.3 *Selection of media and incubation conditions*

Rather than assuming that tryptone soya agar is the appropriate choice, the history of isolates should be taken into account and four or five types of media evaluated with associated incubation conditions of time and temperature. Once or twice a year it may be of value to explore the environment for anaerobes. Tests for thermophiles and psychrophils should also be carried out where environmental conditions may allow selection for those types.

### 8.5.4 *Settle plates*

The position of the lid of the settle plate during exposure should be such that its internal surface is not exposed. Settle plates are often used in the following ways:

1   In Grade A and B environments as described in Annex 1 to the Guide to Good Manufacturing Practice for Medicinal Products (Rules and Guidance for Pharmaceutical Manufacturers and Distributors ('Orange Guide') 1997), settle plates are continuously exposed during machine set-up and processing. A maximum exposure time should be justified to prevent the drying out of agar and loss of growth supporting characteristics of the medium. This may involve preparing plates which have been exposed in laminar flow hoods for worst-case drying conditions, e.g. 1 to > 4 h, and comparing these with undried plates in an uncontrolled area, e.g. in the laboratory or warehouse offices. A 4-h exposure time is often found to be the maximum. Protection from drying out during incubation is important, and use of plastic bags or other methods may help. Settle plates placed in a laminar flow hood should sample positions indicating the conditions to which the product is exposed if possible, as should most environmental monitoring.

There are some differences of opinion about how much monitoring should be allowed to intrude and possibly present a hazard to aseptic processing. Ideally, arrangements should be made to locate settle plates to monitor the Grade A zone at times that do not directly compromise production. A properly argued rationale should be developed for each situation.

2   In Grade C areas used for filling terminally sterilized product and handling some devices, settle plates in challenging locations are exposed at least weekly for 1 h.

For routine positions, a brightly coloured marker permanently on the floor or equipment may help to avoid the plate being trodden on in busy rooms.

3   In Grade D areas, monthly testing with 1 h exposure is often carried out.

### 8.5.5  Surface testing

Contact plates are useful for sampling flat, smooth surfaces and swabs for nooks and crannies. Contact plates tend to give better recovery than swabs when used for flat smooth surfaces. These surfaces should be easy to clean, so routine testing for effectiveness of cleaning using contact plates is not an obvious choice. Cleaning validation however can employ contact plates usefully, provided that any carry-over of antimicrobial residues is first inactivated.

Build-up of contamination on surfaces during operations can be usefully monitored with contact plates, while swabs can effectively check on the cleaning of awkward areas. Contact plates are also useful for testing used garments from aseptic areas. The same test is valuable in the training of new operators when donning sterile garments. After sampling, surfaces should be cleaned to remove nutrient residues.

Swabs need to be moistened before use with a sterile fluid. Saline 0.9%, Ringer's solution or isotonic buffers are more suitable than water in maintaining the viability of micro-organisms during transfer. Likewise, the lag time between sampling and plating swabs should be defined with technical rationale to reduce the risk of dying or proliferation of the organisms captured. The validation of recovery efficiency of swabs and contact plates is beset by questions about relevance of trials to everyday use and the actual need for accuracy, as opposed to repeatability. Suffice it to say, the methods used should be well considered, approved and documented by a microbiologist as suitable for the intended purposes.

In aseptic areas operators are expected to provide finger dabs from each hand at each work session. These agar plates are incubated in the microbiology laboratory and the results evaluated.

### 8.5.6  Air testing – specific notes

Active air sampling provides data that can link the number of colony forming units in the air directly with requirements for room grade [as described in Annex 1 to the Guide to Good Manufacturing Practice for Medicinal Products (Rules and Guidance for Pharmaceutical Manufacturers and Distributors 1997)].

When monitoring Grade A and B zones where limits are very low, aseptic handling skills are required. Out-of-limits counts may be caused by fumbling with agar strips rather than the microbiological quality of the air. The user of air sampling devices should understand the mechanics and dynamics of sampling and entrapment on agar surfaces so that equipment is used properly and results interpreted correctly.

Air particle counting often falls to the environmental microbiologist. With experience, the test can rapidly enable detection of filter failure or equipment break down. If the facility has audible warning output the particle counter is transformed into a valuable problem location and seeking device.

Apart from carrying out the standard test programme of Federal Standard 209E (1992) and BS 5295 (1989), filter faces of laminar flow hoods and room outlets should be scanned for early detection of leaks. Room clean-up rates should also be carried out. It is often forgotten that the Grade A conditions in a laminar flow hood or isolator should be demonstrated during activity, as well as at rest. The wisdom of carrying out intrusive testing when product is being handled should be evaluated. Ideally, arrangements should be made to sample at a time which does not compromise the product, or in some circumstances simulated activity such as broth fills may be used to gather data. If the room is classified at rest it should be shown that it does not reduce by more than one grade when active.

## References

BS 5295 (1989) *Environmental cleanliness in enclosed spaces.* British Standards Institution, London.

BAIRD, R.M. (1999) Sterility assurance; concepts methods and problems. In: RUSSELL, A.D., HUGO, W.B. and AYLIFFE, G.A.J. (eds), *Principles and Practice of Disinfection, Preservation and Sterilization*, 3rd edn. Blackwell, Oxford, pp. 787–800.

DENYER, S.P. and HODGES, N.A. (1998) Sterilization control and sterility assurance. In: HUGO, W.B. and RUSSELL, A.D. (eds), *Pharmaceutical Microbiology*, 6th edn. Blackwell, Oxford, pp. 439–52.

EN 550 (1994) *Sterilization of medical devices – Validation and routine control of ethylene oxide sterilization.* British Standards Institution, London.

EN 552 (1994) *Sterilization of medical devices – Validation and routine control of sterilization by irradiation.* British Standards Institution, London.

EN 1174 (1996, 1997) *Estimation of the population of micro-organisms on product.* British Standards Institution, London.

EN 866 (1997) *Biological systems for testing sterilizers and sterilization processes.* British Standards Institution, London.

*European Pharmacopoeia* (1997) 3rd edn. EP Secretariat, Strasbourg.

Federal Standard 209E (1992) *Cleanroom and workstation requirements, controlled environment.* Washington, DC.

Rules and Guidance for Pharmaceutical Manufacturers and Distributors (1997) *Part Four. Guide to Good Manufacturing Practice for Medicinal Products.* The Stationery Office. ISBN 0-11-321995-4.

# 9

# Endotoxin Testing

ALAN BAINES

*BioWhittaker UK Ltd, Wokingham, Berkshire, RG41 2PL, United Kingdom*

## 9.1  Introduction

### 9.1.1  *Endotoxins and pyrogens*

It is a well-known fact that pyrogenic substances cause a rise in body temperature following intravenous administration. There are many biological and chemical pyrogens, but the most significant pyrogen encountered in the routine production of parenteral and medical devices is Gram-negative bacterial endotoxin. The terms pyrogen and endotoxin are still often interchanged, but it should be emphasized that while all endotoxins are pyrogens, not all pyrogens are endotoxins. During the manufacture of pharmaceuticals and devices, non-endotoxin pyrogens are rarely encountered.

Bacterial endotoxin consists of lipopolysaccharide (LPS) molecules, often attached in the native state to proteins and phospholipids. It forms part of the outer membrane of Gram-negative bacteria and is shed into the surrounding medium. Endotoxin appears in more copious quantities following bacterial lysis due to death or anti-bacterial action, such as sterilization. Endotoxin may, therefore, be present associated with the cell wall of living bacteria or in a free state. The molecular structure of *Salmonella* lipopolysaccharide consists of hydrophobic lipid A attached to a core region which includes three 2-keto-3-deoxyoctonate (KDO) molecules (Figure 9.1). The molecule is amphipathic, causing it to aggregate in aqueous solution. It is also extremely heat stable, so it is essential that endotoxin contamination is avoided *during* manufacture of parenterals and medical devices because it cannot subsequently be removed by autoclaving.

Bacterial endotoxin has a number of physiological effects following intravenous injection including the following: fever; activation of the cytokine system; endothelial cell damage; permeability of vasculature leading to a drop in blood pressure; and intravascular coagulation. In cases where the body is subject to massive doses, such as in the case of Gram-negative bacterial infection, endotoxin release may ultimately cause death from septic shock. While the quantities likely to be administered in a parenteral preparation are orders of magnitude lower than the levels observed in septic

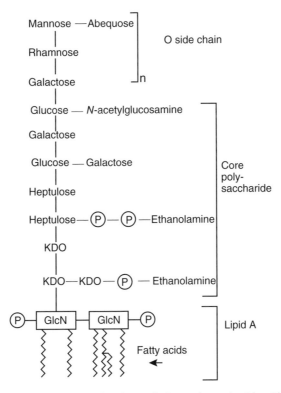

**Figure 9.1** The molecular structure of *Salmonella* lipopolysaccharide. GlcN = glucosamine; KDO = 2-keto-3-deoxyoctonate; P = phosphate.

shock, frequently the patient's immune system may already be compromised by injury or illness, so that the effect of small doses is often magnified.

### 9.1.2 *Regulatory development*

The bacterial endotoxins test (BET) is now a significant part of the workload of most microbiology laboratories involved with the quality control of parenterals and medical devices. Initially, pyrogens were detected by means of the rabbit test, originally developed by Hort and Penfold (1912a,b,c) and subsequently modified prior to inclusion in the *United States Pharmacopoeia* (USP) XII in 1942. The test was later adopted by other pharmacopoeial authorities including the *British Pharmacopoeia* (BP) and the *European Pharmacopoeia* (EP). The rabbit pyrogen test remained the mainstay for nearly 40 years, until the *Limulus* amoebocyte lysate (LAL) test emerged as an alternative during the 1980s. Use of the rabbit test has declined rapidly in the 1990s, and is now limited mainly to older products which have not been converted to the LAL test. The rabbit test will not, therefore, be covered in this chapter and is described elsewhere (Weary, 1996).

The LAL test was developed during the mid-1960s by Levin and Bang (1964) and became a commercial test shortly afterwards. In 1970, Cooper *et al.* published data on the use of the LAL test for the detection of endotoxin in short half-life radio-pharmaceuticals. This early report stimulated its use for the detection of endotoxin in parenteral products and medical devices. Both the regulatory authorities and manufacturers began to consider the LAL test as a viable alternative to the USP rabbit pyrogen test.

The first regulatory acknowledgement of the LAL test came in 1973, when the FDA announced that the LAL reagent was to be licensed as a biological product and that use of the test was restricted to in-process monitoring. Later in 1977, the FDA outlined a conditional approval for the use of LAL as a final release test for licensed biological products and medical devices. At this time the approval did not include non-biological end-product release. Manufacturers were required to submit separate amendments for each product licence, including information on the method used and any supporting data. It was also necessary to show equivalence between the LAL test and the rabbit test to equate the presence of endotoxin *in vitro* with pyrogenicity *in vivo*.

Following the HIMA collaborative study (Dabbah *et al.*, 1980) the FDA approved in draft guidelines for use of the LAL test for all injectable products and medical devices. These stipulated:

- validation procedures for parenteral products including initial qualification of the laboratory and interference tests (inhibition/enhancement testing) for each product;

- the imposition of an endotoxin limit of 0.25 endotoxin units $(EU) ml^{-1}$ for all drug products irrespective of dose; and

- concurrent LAL and rabbit testing for at least three batches of each drug product to demonstrate the absence of any non-endotoxin pyrogenic substances.

Following representations by the industry over concerns about the guidelines (Parenteral Drug Association, 1980), the USP responded with revisions of the BET in 1982 and 1983; these further defined the parameters of the gel-clot test, but still did not address the question of endotoxin limits for individual preparations, a major concern of the pharmaceutical industry. Subsequently, FDA draft guidelines issued in 1983 included the decision to set endotoxin limits based on the maximum human or rabbit dose of the product (whichever was higher), and the adjustment of the endotoxin limit for all drugs except intrathecals from $2.5 EU kg^{-1}$ to $5.0 EU kg^{-1}$. This new guideline encompassed separate guidelines that had previously been issued for biologicals, drugs and medical devices. The USP further revised the BET in 1984 to improve conformation with the procedures described in the 1983 guidelines.

Since 1983 a number of major advances in LAL technology have been made with the introduction of quantitative LAL methods including the end-point chromogenic, the end-point turbidimetric, the kinetic turbidimetric and the kinetic chromogenic assays. The gel-clot test remains the primary method described in both the USP and the EP and the *Japanese Pharmacopoeia* (JP), but in each case the quantitative assays may be used if, 'shown to comply with the requirements for alternative methods', i.e. the manufacturer must show equivalence between the quantitative assay and the BET. In 1987, the FDA released a 'final' guideline for the use of LAL which included a number of minor changes and revisions to the gel-clot test, but more importantly described procedures for the use of quantitative assays. The kinetic chromogenic assay was subsequently covered by two amendments in 1990 and 1991.

Acceptance by European regulatory authorities took longer; the BET first appeared in the EP in 1988 and subsequently in the BP 1989 addendum. Lacking a single European-wide enforcement agency such as the FDA, the provision of European guidelines was delegated to an 'expert' committee which was not entirely successful in producing coherent and workable guidelines, especially for the quantitative methods. Several versions of the 'European guidelines' have been released since 1993, some of which were regarded as unworkable due to a tendency to try to set limits for the quantitative assays which more properly belong to the gel-clot reference test; the last official issue in the 1998 Supplement to the EP is a major improvement. Despite these problems, the acceptance of the LAL test – including the quantitative forms – has become routine with most European authorities.

### 9.1.3 *Introduction to the LAL test*

The LAL assay is used to detect endotoxins associated with Gram-negative bacteria. The lysate, prepared from the circulating amoebocytes of the horseshoe crab (*Limulus polyphemus*), has been shown to be more sensitive to the detection of endo-toxin than the USP rabbit test. There are currently four companies operating in North America and Europe that are licensed by The Food and Drug Administra-tion's Centre for Biologics Evaluation and Research to manufacture LAL. These are BioWhittaker Inc., Associates of Cape Cod, Charles River Endosafe and Haemachem Inc. While they manufacture similar assays, the end-user should evaluate the differ-ent lysates for their method of choice, as the characteristics of each manufacturers' lysates can differ considerably.

The use of LAL for the detection of endotoxin evolved from the observation by Bang (1956) that Gram-negative infection of *Limulus polyphemus* resulted in fatal intravascular coagulation. Levin and Bang (1964) later demonstrated that this clotting was a result of the action between endotoxin and a clottable protein in the circulating amoebocytes of *Limulus*. Following the development of a suitable anticoagulant for *Limulus* blood, Levin and Bang prepared a lysate from washed amoebocytes which was an extremely sensitive indicator of the presence of endotoxin. Solum (1970, 1973) and Young *et al.* (1972) purified and characterized the clottable protein from LAL, and have shown the reaction with endotoxin to be enzymatic.

There are four LAL methodologies currently licensed by the FDA and recognized in the 1998 EP. The first, referred to as the gel-clot method, is based on the fact that LAL clots in the presence of endotoxin. The kinetic turbidimetric is a quantitative LAL method which utilizes the rate of gelation to determine endotoxin content. The third and fourth are chromogenic methodologies employing a synthetic chromogenic substrate which, in the presence of LAL and endotoxin, produces a yellow colour that is linearly related to the endotoxin concentration.

The gel-clot test was the first to be developed and is still extensively used today. This method provided a simple bench-top limit test that is well suited to applications that do not necessarily require quantitation, such as product release testing. However, lack of quantitative information and problems with interference from test substances prompted the development of new forms of the LAL test, which are now in routine use in the pharmaceutical and medical device industries. The end-point chromogenic method, the kinetic turbidimetric method and the kinetic chromogenic method are all improvements on the basic gelation test. The choice of method for the indi-vidual laboratory will depend on the applications, test volume and type of product. Each method is described below, together with the necessary equipment to carry out the test.

The LAL test utilizes the basic immune response of the horseshoe crab to Gram-negative bacterial invasion. The materials contained in horseshoe crab amoebocytes comprise the various proteins, factors, co-factors and ions that interact to initiate coagulation. Gram-negative bacterial endotoxin catalyses the activation of a proenzyme in the *Limulus* amoebocyte lysate (Levin and Bang, 1968). The initial rate of activation is determined by the concentration of endotoxin present. The activated enzyme (coagulase) hydrolyses specific bonds within a clotting protein (coagulogen) also present in *Limulus* amoebocyte lysate. Once hydrolysed, the resultant coagulin self-associates and forms a gelatinous clot.

One problem found by all users of the LAL test is the frequent interference with the assay from the test compound. The only published study (Guilfoyle and Munson, 1982) demonstrated that of 587 compounds tested, 77% interfered with the LAL assay to some degree. Given the continual rise in the number of biopharmaceutical products being brought to market, that figure is probably higher still today. Figure 9.2 shows the involvement of a number of factors in the 'cascade' reaction which results in a gel-clot; interference may occur at several points in this sequence. Actions that the user can take to overcome interference are discussed later in the chapter.

The problems highlighted above, coupled with a desire for endotoxin quantitation, led to research into alternative methods to the gel-clot test; the first was the chromogenic end-point reaction. As clot formation is the result of enzymatic cleavage of a protein, it was a logical step to look for an alternative substrate for the enzyme responsible for the generation of the clotting protein. Protein sequencing indicated the structure of the cleavage site on the coagulogen. This information led to synthesis of a short-chain peptide which would mimic the sequence of amino acids prior to the cleavage site and conjugation to this peptide of a suitable chromophore. The chromophore normally used is *para*-nitroaniline (PNA), which is colourless when linked to the peptide, turning to yellow when dissociated. The yellow colour can easily be measured by a spectrophotometer at 405 nm. As the quantity of PNA released is directly proportional to the amount of enzyme liberated during the cascade, and the

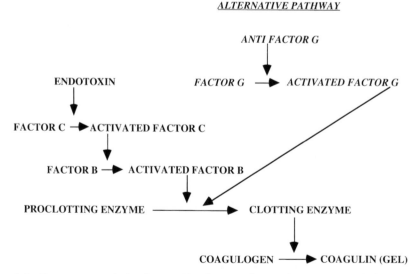

**Figure 9.2**  Enzyme cascade leading to *Limulus* amoebocyte lysate (LAL) gel-clot.

amount of enzyme is directly proportional to the amount of endotoxin in the reaction vessel, the requirements for quantitation are satisfied.

The chromogenic reaction avoided the need to rely on the clotting phase of the reaction sequence, and the end-point chromogenic test provides limited quantitation. Subsequently, kinetic methods were developed which modified the basic biochemistry to enable a wider range of quantitation and improved sensitivity. The kinetic turbidimetric assay uses the same basic reaction sequence as the gel-clot test, but in this case the reaction is not allowed to proceed to conclusion. The kinetic chromogenic assay was a natural development of the end-point method, and this method is technically and practically the most successful LAL assay in use today.

## Determination of endotoxin limits

Before the testing of final products, bulk intermediates or medical devices can begin, it is necessary to obtain or calculate the endotoxin limit for the preparation or item in question. The endotoxin limit may, for some products, be obtained from the USP or EP monograph. If there is no predetermined limit available from the pharmacopoeia, then it must be calculated from the maximum human dose. A number of different terms – endotoxin limit (EL), maximum allowable endotoxin concentration (MAEC), endotoxin release limit (ERL) or endotoxin limit concentration (ELC) – are used by the different regulatory bodies, but in essence they all mean the same thing. They indicate the maximum amount of endotoxin that should not be met or exceeded in a pharmaceutical, or be present on a medical device. If there is no monograph available, then the endotoxin limit may be determined from the formula:

Endotoxin limit $= K/M$,

where K is a constant equal to 5 EU or IU per kg of body weight, and $M$ is equal to the maximum human dose administered per kg per hour. Another figure that is often quoted is an endotoxin limit of 350 EU per hour for the 'average' 70-kg man. In practice, the $5 \, EU \, kg^{-1}$ figure is much more useful as it will allow for the correct calculation of dose for all size and age ranges. The laboratory should be cognisant of the fact that the endotoxin limit is as indicated per hour, when making such calculations. It can quickly be seen that the allowable concentration of endotoxin per millilitre in an injectable preparation will vary considerably with the volume administered within the hour. A single 1-ml injection could theoretically contain $349 \, EU \, ml^{-1}$ and still pass the BET, whereas a 1-litre infusion would need to contain less than $0.349 \, EU \, ml^{-1}$.

## MVD and MVC calculations

The maximum valid dilution (MVD) or minimum valid concentration (MVC) are calculated figures that indicate the degree to which a product may be diluted to overcome interference, before the effect of the dilution exceeds the ability of the LAL test method being used to detect endogenous endotoxin in the original preparation. The term MVD is normally applied to preparations already in a liquid form where the dose is administered per millilitre; for example, a single 2-ml injection and the endotoxin limit is expressed as $EU \, ml^{-1}$. The term MVC is applied to those preparations where the endotoxin limit is expressed as $EU \, mg^{-1}$ and the dose is expressed as mg/kg of body weight.

The determination of the MVD or MVC is dependent on the sensitivity of the lysate used, which is represented in the equations below by $\lambda$ (see section 9.2.2). In the case of gel-clot lysates, $\lambda$ is the labelled sensitivity of the lysate, e.g. 0.06 EU ml$^{-1}$, whereas for the quantitative methods $\lambda$ represents the value of the lowest standard used in the assay for the creation of the standard curve, e.g. 0.005 EU ml$^{-1}$ for the kinetic chromogenic method. The more sensitive the lysate or method, the higher the MVD or lower the MVC value will be.

The MVC is calculated from the formula:

$$MVC = \frac{\lambda M}{K}$$

where $\lambda$ is the sensitivity of the lysate or the value of the lowest standard for quantitative assays, K is a constant equal to 5.0 EU kg$^{-1}$, and $M$ is the maximum human dose.

The MVD is calculated from the formula:

$$MVD = \frac{c \times K}{\lambda \times M}$$

where $c$ is the concentration of drug, and K and $M$ remain the same.

Thus, for example, an insulin injection of potency 100 units ml$^{-1}$, a maximum dose of 2 units kg$^{-1}$ and a lysate sensitivity of 0.125 EU units ml$^{-1}$ would have:

$$MVD = \frac{100 \times 5}{0.125 \times 2} = 1 : 2000.$$

An MVD value can be calculated from an MVC value:

$$MVD = \frac{potency\ of\ product}{MVC}$$

### Medical devices

Levels of endotoxin on medical devices are obtained by an extraction procedure. This involves soaking a number of devices in an extracting fluid, normally LAL reagent water. A limit of 20 EU per device was set in the 1997 USP addendum, so the maximum permissible endotoxin concentration in the extracting fluid (endotoxin release limit, ERL) can be calculated thus:

$$ERL = \frac{K \times N}{V}$$

where K is 20 EU per device, $N$ is the number of devices, and $V$ is the total volume of the extraction solution.

## 9.2  The gel-clot method

Although variants of the gel-clot test exist (e.g. the slide-spot, capillary and micro-plate methods), these deviations from the standard test are usually more subjective and have no audit trail, as they are not supported by any of the manufacturer's quality control procedures. While they are allowed by certain pharmacopoeias, provided that validation studies have been carried out, they may disappear as a result of

**Table 9.1**  Equipment required for gel-clot and chromogenic end-point methods

| Gel-clot method | Chromogenic end-point method |
|---|---|
| • Gel-clot lysate with matched control standard endotoxin (CSE) | • Gel-clot lysate with matched control standard endotoxin (CSE) |
| • Dry heat block (preferred) or non-recirculating water bath at $37 \pm 0.5°$C | • Dry heat block with insert for microplate (preferred) or non-recirculating water bath at $37 \pm 0.5°$C |
| • Vortex mixer | • Vortex mixer |
| • 100 µl pipette | • Variable micropipette: range 50–100 µl |
| • Depyrogenated glass reaction tubes | • Eight-channel micropipette: range 50–100 µl |
| • Depyrogenated glass dilution tubes | • Micropipette: fixed 10 µl volume |
| • Either variable volume 1 ml or 5 ml pipettes, or apyrogenic serological pipettes for reconstitution of reagents | • Apyrogenic 96-well microtitre plates |
| | • Depyrogenated glass dilution tubes |
| | • Either variable volume 1 ml or 5 ml pipettes, or apyrogenic serological pipettes for reconstitution of reagents |
| | • Graph plotting software an advantage |

moves to harmonize the BET on a global basis. The use of such variants will not, therefore, be dealt with in this chapter.

The items of equipment required are shown in Table 9.1. Most standard laboratory suppliers can provide the heating unit, vortex mixer and pipettes. The user is however, advised to buy directly from a lysate manufacturer those accessories where absence of endogenous endotoxin is essential, e.g. pipettes. Use of a dry heat block is recommended as it provides a more accurate, consistent and easily controlled heat source. Recirculating water baths are unsuitable as the vibration from the pump may result in gel fragmentation or failure to form properly. As a recirculating unit cannot be used, the bath volume must be relatively small to avoid temperature fluctuations.

### 9.2.1  *Test principle and procedure*

Following the manufacturer's instructions, accurate volumes of sample or standards prepared from the CSE (typically 100 µl) are pipetted into $10 \times 75$ mm depyrogenated glass tubes. The kit insert will normally describe the correct dilution schemes; these usually comprise two-fold dilutions of an endotoxin standard, dilutions of the test sample, and LAL reagent water to serve as a negative control. After addition of 100 µl reconstituted lysate, all tubes are transferred to a 37°C water bath or heating block and incubated for 1 h. Depending on the type of assay to be performed, the addition of positive product controls (PPCs) may also be required; these should be added prior to addition of the lysate. After incubation, the tubes must be removed and inverted by 180° in a single, smooth motion to determine if a solid clot has formed at the bottom. Inevitably there will be occasions where the clot initially appears firm but then breaks away from the base of the tube; such an occurrence should always be judged as a negative result. The exact time that the tube is held in an inverted position is somewhat subjective, but should be no more than 2 s; more than one inversion of the tube is not required and invalidates the test. Each tube where a firm clot has formed is recorded as positive and any other result as negative. Due to the nature of the test it is not possible to legislate for slight variations in the incubation temperature, inversion time, etc. which can make the difference between a borderline positive and a negative result. However, adherence to the following precautions promotes reproducibility:

1 All materials coming into contact with specimen or test material must be pyrogen-free. Materials may be rendered pyrogen-free by heating at 180°C for 4 h, 250°C for 30 min, or any other validated method.

2 Careful technique must be used to avoid contamination with endogenous endotoxin.

3 Strict adherence to incubation time and temperature is required.

4 The test specimen should be in the pH range of 7.0 to 8.0. If required, pH may be adjusted with pyrogen-free acid or base.

### 9.2.2  *Gel-clot lysate sensitivity*

In order to assign significance to positive or negative gel-clot results, it is necessary to determine the minimum concentration of endotoxin which causes gelation of the lysate, known as the lysate sensitivity. Each vial of LAL is labelled with the lysate sensitivity obtained by the manufacturer using the FDA Reference Standard Endotoxin (RSE), and is expressed in endotoxin units (EU) or international units (IU). Since the adoption of Biological Reference Preparation No. 3 (BRP-3) by the EP it is now a fact that $1\,EU = 1\,IU$; the user may now, therefore, use these terms interchangeably. Each user should re-verify the labelled lysate sensitivity as part of initial quality control. This is a regulatory requirement in both FDA and EP guidelines for new technicians and each new lot of lysate used in the laboratory. Using an endotoxin standard, normally known as the Control Standard Endotoxin (CSE) whose potency is known, serial two-fold dilutions of the CSE should be prepared. These dilutions should bracket the labelled lysate sensitivity. Usually each dilution, as well as a negative water control, is assayed in quadruplicate. Incubation and interpretation of results are as described in section 9.2.1. The end-point dilution is determined as the last dilution of endotoxin which still yields a positive result (Table 9.2).

**Table 9.2**  Specimen assay results for the gel-clot method

Endotoxin dilution (EU ml$^{-1}$)

| Replicate | 0.5 | 0.25 | 0.125 | 0.06 | 0.03 | 0 (water) | End-point |
|-----------|-----|------|-------|------|------|-----------|-----------|
| 1 | + | + | + | − | − | − | 0.125 |
| 2 | + | + | + | − | − | − | 0.125 |
| 3 | + | + | + | + | − | − | 0.06 |
| 4 | + | + | + | − | − | − | 0.125 |

The lysate sensitivity is calculated by determining the geometric mean of the end-point. Each end-point value is converted to $\log_{10}$. The individual $\log_{10}$ values are averaged and the lysate sensitivity is taken as the antilog$_{10}$ of this average $\log_{10}$ value; from Table 9.2 this would be calculated thus:

|  | End-point (EU ml$^{-1}$) | Log$_{10}$ end-point |
|---|---|---|
| Row 1 | 0.125 | −0.903 |
| Row 2 | 0.125 | −0.903 |
| Row 3 | 0.06 | −1.222 |
| Row 4 | 0.125 | −0.903 |
|  | mean $\quad$ = | −0.983 |
|  | antilog$_{10}$ mean = | 0.10 EU ml$^{-1}$ |

The acceptable variation range is from one-half to twice the labelled lysate sensitivity. In the example above, geometric mean values exactly equal to 0.06 or 0.25 are within the allowed limits and are therefore acceptable. The laboratory should then always use the claimed sensitivity of the lysate (label claim) in any further studies or calculations such as MVD determinations.

### 9.2.3  *Product interference*

The LAL reaction is enzyme-mediated and so has an optimal pH range and specific salt and divalent cation requirements. Occasionally test samples may alter these optimal conditions such that the lysate is no longer sensitive to endotoxin. Therefore, negative results with samples which inhibit the LAL test do not necessarily indicate the absence of endotoxin. Occasionally, there may also be factors that cause a false-positive reaction or 'enhancement'. Such results could be due to enzymes within the preparation acting directly to cleave the coagulogen, or be due to glucans or LAL-reactive material (LAL-RM) as these are often termed.

Initially, each type of sample should be screened for product inhibition and enhancement. A series of two-fold dilutions of standard endotoxin in water should be prepared in parallel with a similar series of endotoxin dilutions using a sample as diluent. Each series should be assayed in parallel using standard procedures. At the end of the incubation period, positive and negative results should be recorded and the geometric mean end-point calculated for both series of endotoxin dilutions. Products are said to be free of product inhibition if the geometric mean end-point of endotoxin in the product and the geometric mean end-point of a similar series of endotoxin in water are within one-half to twice the lysate sensitivity. Specimen results are shown in Table 9.3.

The easiest way to overcome product inhibition is through dilution. This initial dilution factor must be taken into account when calculating the total endotoxin concentration in a test sample. As a quick screen to determine a non-inhibitory dilution, a series of increasing product dilutions should be prepared containing an endotoxin spike equal in concentration to twice the lysate sensitivity. Each spiked product dilution should be assayed using standard procedures. Positive results indicate when product inhibition has been overcome. Products which are extremely acidic or basic may require pH adjustment to neutrality, as well as dilution, in order

**Table 9.3**  Specimen results for product inhibition testing

|  |  | Endotoxin dilution (EU ml$^{-1}$) | | | | |
|---|---|---|---|---|---|---|
|  | Replicate | 0.5 | 0.25 | 0.125 | 0.06 | 0.03 |
| Endotoxin in water | 1 | + | + | + | − | − |
|  | 2 | + | + | + | − | − |
|  | 3 | + | + | + | + | − |
|  | 4 | + | + | + | − | − |
| Geometric end-point $= 0.10$ EU ml$^{-1}$ (see Table 9.2 and text) | | | | | | |
| Endotoxin in product A | 1 | + | + | − | − | − |
|  | 2 | + | + | + | − | − |
|  | 3 | + | + | + | − | − |
|  | 4 | + | + | + | − | − |
| Geometric end-point $= 0.14$ EU ml$^{-1}$, non-inhibitory | | | | | | |
| Endotoxin in product B | 1 | + | − | − | − | − |
|  | 2 | + | − | − | − | − |
|  | 3 | + | − | − | − | − |
|  | 4 | + | − | − | − | − |
| Geometric mean end-point $= 0.50$ EU ml$^{-1}$, inhibitory | | | | | | |

to overcome product inhibition. Thus, 0.1 M NaOH or HCl prepared by dilution of a 1 M concentration of acid or base with LAL reagent water is suitable. A better alternative is the use of a LAL reagent grade Tris buffer that is available from some LAL suppliers.

The laboratory should avoid testing at the limits of sensitivity, as defined by the MVD, if at all possible so that, should product inhibition occur, a re-test is still possible. The gel-clot test will always have a degree of subjectivity that is not present in the quantitative methods. This does not, however, detract from its value as a pharmacopoeial limit test. Each laboratory should establish its own test limits which should be consistent across the whole product range. In the author's experience a value equal to one-half MVD is a suitable choice. This internal limit will prevent the necessity for re-tests that a stricter internal limit may require, while allowing for a safety margin for the release of product.

Some compounds may induce or give false-positive or false-negative results in a gel-clot test, e.g. blood products, polynucleotides, solutions containing heavy metals or surfactants, or those of high ionic strength or osmolarity. Non-specific inhibitors in blood products may be removed by diluting the test sample 1 : 3 in pyrogen-free water and heating to 100°C for 10 min.

## 9.3  The chromogenic end-point method

The chromogenic end-point method offers a fast, relatively low-cost method which allows quantitation and frequently overcomes interference problems that occur with the gel-clot or kinetic turbidimetric methods. It is the most rapid LAL test, producing results in around 20 min, but is also the most demanding in terms of operator technique. The method is best suited to assays where either a small number of products are tested or the assay is only run infrequently. It is an ideal solution for those laboratories where one or two products cause validation problems with the gel-clot test, but the cost of solving the problem via a kinetic assay system is not justified. It is also commonly used for testing serum samples for clinical studies where additional cost of an automated system is not warranted.

Gram-negative bacterial endotoxin catalyses the activation of a proenzyme in the LAL test (Dabbah *et al.*, 1980). The initial rate of activation is determined by the concentration of endotoxin present. The activated enzyme catalyses the splitting of PNA from the colourless substrate Ac-Ile-Glu-Ala-Arg-pNA. The PNA released is measured spectrophotometrically at 405 nm after the reaction is stopped with an appropriate reagent. The correlation between the absorbance and the endotoxin concentration is linear in the 0.1 to 1.0 EU ml$^{-1}$ range. The concentration of sample endotoxin is calculated from the absorbance values of solutions containing known amounts of endotoxin standard.

The equipment required is listed in Table 9.1. This assay can be run in depyrogenated glass tubes in a water bath or dry heating block, should suitable arrangements for heating the microtitre plate not be available. Due to the short incubation times and the need to maintain both the absolute temperature and the temperature uniformity across the microtitre plate, use of other heating devices such as warm air incubators is not recommended. Most dry heat block suppliers can provide a blank for machining, should a custom-made microtitre plate insert not be available. If using glass tubes, the contents can be transferred after the reaction has been stopped, either to a microtitre plate or a spectrophotometer for reading of the results. The reaction volume is usually around 300 µl which may necessitate the use of a 'micro-cell' or doubling of reagent volumes. Diluting the final volume to fit a standard spectrometer cell is not recommended, as the slope of the calibration line will be reduced giving poor differentiation between results.

### 9.3.1 *Test principle and procedure*

The reagent volumes and incubation times given below are typical of current assays, but will vary according to the kit manufacturer. The precautions necessary to avoid erroneous results and enhance reproducibility are as described for the gel-clot test (see section 9.2.1).

Using an apyrogenic microtitre plate, 50 µl of sample or standard should be pipetted into the plate. If positive product controls are required, a 10 µl pipette may be used to add sufficient endotoxin to the appropriate wells to give a concentration of 4λ (four times the concentration of the lowest standard). After pre-warming the plate for 10 min, 50 µl of lysate is added using a multichannel pipette and the plate incubated for 10 min. An eight-channel micropipette is used to add 100 µl of reconstituted, pre-warmed chromogenic substrate followed by incubation for a further 6 min. Acetic acid (10%, 100 µl) may be added to stop the reaction, although sodium dodecyl sulphate (SDS, 10%) is an alternative should the acetic acid be incompatible with the product. The mean absorbance of the blank should be subtracted from the mean absorbance values of the standards and samples to calculate mean absorbance changes, and these plotted against endotoxin concentration; under normal conditions, the plot is linear between

**Table 9.4**  Specimen data for an end-point chromogenic *Limulus* amoebocyte lysate (LAL) test

| Well | Endotoxin source | Absorbance (A) | Mean A | Mean ΔA (sample A – blank A) |
|------|------------------|----------------|--------|------------------------------|
| 1  | LAL reagent | 0.080 | | |
| 2  | Water (blank) | 0.084 | 0.082 | |
| 3  | 0.1 EU ml$^{-1}$ standard | 0.157 | | |
| 4  | 0.1 EU ml$^{-1}$ standard | 0.182 | 0.170 | 0.088 |
| 5  | 0.25 EU ml$^{-1}$ standard | 0.309 | | |
| 6  | 0.25 EU ml$^{-1}$ standard | 0.325 | 0.317 | 0.235 |
| 7  | 0.50 EU ml$^{-1}$ standard | 0.570 | | |
| 8  | 0.50 EU ml$^{-1}$ standard | 0.557 | 0.564 | 0.482 |
| 9  | 1.0 EU ml$^{-1}$ standard | 1.052 | | |
| 10 | 1.0 EU ml$^{-1}$ standard | 1.012 | 1.032 | 0.950 |
| 11 | Product # 1 | 0.372 | | |
| 12 | Product # 1 | 0.392 | 0.382 | 0.300 |
| 13 | Product # 2 | 0.916 | | |
| 14 | Product # 2 | 0.912 | 0.914 | 0.832 |

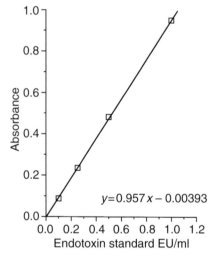

$$y = 0.957x - 0.00393$$

**Figure 9.3**  Calibration plot for chromogenic end-point assay (taken from data in Table 9.4).

0.1 to 1.0 EU ml$^{-1}$ endotoxin. The regulatory bodies require a linear regression analysis of the plotted data to show a correlation coefficient ('*r*' value) of $\geq 0.98$. The line should also have a significant slope. Specimen data are shown in Table 9.4; these are plotted in Figure 9.3.

The endotoxin concentrations may be read directly from the plot, or calculated from the equation representing the line. Most graph-plotting packages are capable of calculating such equations, but the detailed method is also shown in Chapter 14 (using data from Table 9.4). If the calculated endotoxin concentration is greater than 1.0 EU ml$^{-1}$, it is necessary to dilute the sample five-fold in LAL reagent water, re-test, and multiply the new value by 5.

### 9.3.2 *Performance characteristics*

In addition to a correlation coefficient of $\geq 0.98$ mentioned above, it should be possible to achieve, on a routine basis, coefficients of variation (standard deviation as a percentage of the mean) for replicate samples of less than 10%; with experience, values of 3–4% should be attainable.

Product inhibition or enhancement occurs when substances in the test sample interfere with the LAL reaction. In the chromogenic assay, inhibition results in a lower, final $\Delta$ absorbance, which leads to an artificially low calculated endotoxin concentration; the converse occurs with enhancement. The lack of product inhibition or enhancement should be determined for each specific sample, either undiluted or at an appropriate dilution. In order to verify the lack of product interference, an aliquot of test sample (or a dilution thereof) is spiked with a known amount of endotoxin ($0.4\,\mathrm{EU\,ml}^{-1}$, i.e. 4 × concentration of the lowest standard; FDA Guidelines). This can be achieved simply by diluting an appropriate volume of endotoxin stock solution using the test sample (or dilution) as diluent, followed by vigorous vortexing for 1 min. The spiked solution is assayed along with the unspiked sample and their respective endotoxin concentrations are determined. The difference between these two calculated endotoxin values should equal the known concentration of the spike $\pm 25\%$. If inhibition or enhancement is identified, the sample may require further dilution until the interference is overcome. Specimen data illustrating interference are shown in Table 9.5.

Initially, 10-fold dilutions of test sample may be screened for product interference. Once the approximate non-interfering value is determined, the exact dilution can be found by testing two-fold dilutions in this region.

In the end-point chromogenic assay, significantly coloured samples may require special attention. It should be noted that some turn yellow in acid environments, e.g. tissue culture media. A quick test to determine if a product's intrinsic colour is sufficient to be of concern is to construct a sample blank by combining 50 μl of sample, 150 μl of LAL reagent water and 100 μl stop reagent without incubation. If the 405 nm absorbance of this blank, when read against LAL reagent water, exceeds 0.5 the background colour is significant, and the sample should be diluted and re-assayed. The

**Table 9.5** Data illustrating product interference with end-point chromogenic, and kinetic turbidimetric *Limulus* amoebocyte lysate (LAL) tests

| Sample dilution | End-point chromogenic assay | | | Kinetic turbidimetric assay | |
|---|---|---|---|---|---|
| | Endotoxin concentration recorded $(\mathrm{EU\,ml}^{-1})$ | | | | |
| | Unspiked | Spiked $(0.4\,\mathrm{EU\,ml}^{-1})$ | Difference | Sample dilution | Endotoxin recovered $(\mathrm{EU\,ml}^{-1})$ (spike $0.1\,\mathrm{EU\,ml}^{-1}$) |
| 1/10 | 0.18 | 0.28 | 0.10 | 1/10 | 0.015 |
| | Conclusion: product inhibitory | | | Conclusion: inhibitory | |
| 1/20 | 0.09 | 0.36 | 0.27 | 1/20 | 0.042 |
| | Conclusion: product inhibitory | | | Conclusion: inhibitory | |
| 1/40 | 0.04 | 0.44 | 0.40 | 1/40 | 0.110 |
| | Conclusion: product non-inhibitory | | | Conclusion: non-inhibitory | |

dilution factor is then used in the final calculations for determining the concentration of endotoxin. In practice, this sample blank is assayed alongside the standards, product and water reagent blank in the normal way and the $\Delta$ absorbance of the sample is calculated by subtracting the values for both the water reagent blank and the sample blank; only the LAL reagent water blank is used to calculate the $\Delta$ absorbance for endotoxin standards and non-coloured products.

By extending the first incubation it is possible to measure concentrations of endotoxin below $0.10\,\mathrm{EU\,ml^{-1}}$. Typically, an initial 30-min incubation would permit detection in the range 0.01 to $0.10\,\mathrm{EU\,ml^{-1}}$, although incubation times will vary according to the manufacturer. It is important to determine the linearity of the test at this higher sensitivity and run standards containing an appropriately lower concentration of endotoxin.

The degree of inhibition or enhancement will be dependent upon the concentration of the product. If several concentrations of a product are to be assayed, performance characteristics should be established for each independently. Patterns of inhibition or enhancement different from those seen with the traditional LAL gelation test may be found. If necessary, sample pH should be adjusted within the range of 7.0 to 8.0 using pyrogen-free sodium hydroxide or hydrochloric acid or Tris buffer to overcome inhibition.

## 9.4   The kinetic turbidimetric assay

Teller and Kelly (1979) first demonstrated that the endotoxin activation of lysate could be monitored photometrically. If a sample is mixed with reconstituted LAL reagent and automatically monitored for the appearance of turbidity, the time required to reach a predetermined absorbance value (reaction time) is inversely proportional to the amount of endotoxin present. Thus, the concentration of endotoxin in unknown samples can be calculated from a standard curve. Both the kinetic turbidimetric and the kinetic chromogenic assays require a heated ELISA plate reader, or alternatively a heated tube reader. The ease of use, adjunct technology such as multipipettes and the high sample capacity make the plate readers a better option for all but a few laboratories. Whichever system is chosen, the software should be capable of providing a log-log curve fit to the data. Choosing a system from one of the main lysate manufacturers is recommended as they usually have specialized software for LAL testing, which makes the test procedure and generation of results easier.

### 9.4.1   *Reagent preparation*

In order to calculate endotoxin concentrations in unknown samples, each kinetic turbidimetric LAL assay must be referenced to a valid standard curve. Owing to the potentially large concentration range over which endotoxin values can be determined, it is possible to adjust the quantitative range of any given assay by adjusting the concentration of endotoxin standards used to generate the standard curve.

The day-to-day operating range of the kinetic turbidimetric assay will depend on both the lysate and the software used in the assay. The laboratory should ascertain the recommended operating range from the lysate manufacturer prior to choosing a standard range. The turbidimetric assay does suffer from a more pronounced 'bow'

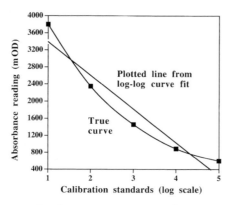

**Figure 9.4**   Over-estimation of endotoxin concentration due to use of log-log curve fit.

in the standard curve than its chromogenic counterpart (Figure 9.4), which may lead the individual user to truncate the standard curve to one or two log, e.g. 0.01 to 1.0, depending on specific product requirements. The bow results in over-estimation of the endotoxin concentration for samples in the middle of the range because the traditional curve-fitting program represents as a straight line data which should be represented by a curve (Figure 9.4); this poses a particular problem in validation studies because spike recoveries may be high.

Data indicate that truncating a turbidimetric LAL standard curve may improve the accuracy of predicted endotoxin values for test samples when using a basic log-log curve-fitting routine. A recent development by BioWhittaker, the Powercurve®, addresses this problem and restores the operating range from 0.01 to 10 or even 0.01 to 100 EU ml$^{-1}$. Laboratories should be familiar with the FDA interim guidance requirements for kinetic LAL techniques prior to establishing a turbidimetric LAL standard curve range to be used for routine testing of product samples.

The initial solution containing 10 EU ml$^{-1}$ endotoxin is prepared by adding 0.1 ml of the 100 EU ml$^{-1}$ endotoxin stock into 0.9 ml of LAL reagent water. This solution should be vigorously vortexed for at least 1 min before proceeding. Successive 10-fold dilutions are similarly prepared down to 0.01 EU ml$^{-1}$.

### 9.4.2   *Test procedures*

As with the other methods, the kinetic turbidimetric assays are performed with the standard endotoxin solutions, product samples and LAL reagent water blanks run in duplicate. The assay relies on the change in the measured absorbance due to light scattering as the small clot fragments start to develop. A kinetic plate reader will record the new optical density reading each time the plate is read, which is usually at 1-min intervals. As the assay measures the change in optical density, it is possible to account for initial colour in the sample. The instrument/software setting should be programmed for a kinetic turbidimetric LAL test according to the manufacturer's instructions. The reader must be located in an area free of excessive vibration (e.g. centrifuges, shakers, etc.) while the test is run, as it is just as susceptible to vibration as the gel-clot test. Due to the very low signal-to-noise ratio seen with the measurement of the absorbance change, care must be taken to avoid bubbles. It is advisable to reconstitute most turbidimetric lysates at least 10 min before the start of the reading phase of the assay to allow dispersion of bubbles or foam.

### 9.4.3  *Performance characteristics*

The linearity of the standard curve within the concentration range used to determine endotoxin values should be verified. No less than three endotoxin standards, spanning the desired concentration range, and a LAL reagent water blank should be assayed in quadruplicate according to the test parameters of an initial qualification assay. Additional standards should be included to bracket each log interval over the range of the standard curve. The absolute value of the coefficient of correlation, 'r', of the calculated standard curve should be between $-0.980$ and $-1.000$ (the kinetic assays have a negative slope) and the coefficient of variation of replicate sample determinations should again be $<10\%$, with values of 3–4% being attained with practice.

In the turbidimetric LAL assay, any product inhibition results in a longer reaction time, and artificially low endotoxin concentrations are recorded. The lack of product inhibition should be determined in a similar manner to that described in section 9.3.2. It is recommended that the endotoxin spike (positive product control, PPC) produces a final endotoxin concentration in the sample of $0.1\,\mathrm{EU\,ml}^{-1}$. For samples which may contain a background endotoxin level $>0.1\,\mathrm{EU\,ml}^{-1}$, the spike should result in a final endotoxin concentration of $1.0\,\mathrm{EU\,ml}^{-1}$. A spiked aliquot of the test sample (or dilution) at a concentration of $0.1\,\mathrm{EU\,ml}^{-1}$ may be prepared by addition of 50 µl of a $10\,\mathrm{EU\,ml}^{-1}$ solution of endotoxin in LAL reagent water to 4.95 ml of test sample (or dilution); this solution should be vigorously vortexed for 1 min prior to use. When preparing the plate it is normal to pipette two sets of duplicate wells for each sample that requires a PPC. The first set is used to measure the endogenous endotoxin, while the second set has the PPC added. The mean value for the sample wells is then subtracted from that for the PPC wells to give an accurate recovery value for the PPC. This value should be within $\pm 50\%$ under FDA guidelines. The EP (1998) allows a wider remit, with the allowed range being 50% to 200% of the original value.

Specimen data illustrating interference are shown in Table 9.5. If the test sample (or dilution) is found to be inhibitory to the turbidimetric LAL reaction, the sample may require further dilution until the inhibition is overcome. The degree of inhibition or enhancement will be dependent upon the concentration of product. If several concentrations of the same product are to be assayed, it is necessary to establish performance characteristics for each independently. Again, patterns of inhibition or enhancement different from those seen with the gel-clot test may be found, and pH adjustment may be necessary as described in section 9.3.2. Samples that possess significant turbidity may require clarification by centrifugation, filtration or dilution prior to testing.

### 9.5  **The kinetic chromogenic assay**

The kinetic chromogenic assay was first launched in 1989 and rapidly became one of the highest volume LAL assays in use today. Its success was due to the ease-of-use of

**Table 9.6** Assay parameters for chromogenic and turbidimetric kinetic *Limulus* amoebocyte lysate (LAL) assays

|  | Kinetic turbidimetric assay | Kinetic chromogenic assay |
|---|---|---|
| Optical filter (nm) | 340 | 405 |
| Absorbance increase (mOD)* | 30 | 200 |
| Number of repeat readings | 100 | 40 |
| Reading interval (s) | 60 | 150 |

* $10^{-3}$ OD units.

a single-step method and the lower levels of interference seen in the chromogenic reaction. With the end-point method, the lysate and substrate are supplied separately and have to be added individually by the operator. The combination of short incubation times and multiple reagent additions of the end-point method, while yielding rapid results, compromise reproducibility and accuracy is more difficult to achieve. In the kinetic chromogenic test, the lysate and substrate have been co-lyophilized so that following the preparation of the samples and standards, a single addition of reagents is necessary. The kinetic chromogenic method uses the same detection method and calculation software as its turbidimetric counterpart. The main differences, with examples of the likely instrument settings, are highlighted in Table 9.6.

It will be seen from Table 9.6 that the signal-to-noise ratio for the kinetic chromogenic method is significantly better than that for the kinetic turbidimetric method. This is one of the reasons why the kinetic chromogenic method is better able to cope with preparations that exhibit turbidity or slight precipitation. Kinetic chromogenic assays are nearly always conducted using an ELISA plate reader, although it is possible to run the assay on some types of tube reader.

### 9.5.1 *Test procedure*

As with the kinetic turbidimetric method, the assay of unknown samples is dependent on the creation of a standard curve from the Control Standard Endotoxin. Owing to the large concentration range over which endotoxin values can be determined, it is possible to alter the quantitative range of any assay by adjusting the concentration of endotoxin standards used to generate the standard curve. Typically, 10-fold dilutions of endotoxin in LAL reagent water are prepared at concentrations between 5.0 and 0.005 EU ml$^{-1}$; vigorous vortexing for at least 1 min is required at each step. Throughout the assay, the ELISA reader continuously monitors the absorbance at 405 nm of each well of the micro-plate. Using the initial absorbance reading of each well as its own blank, the reader determines the time required (reaction time) for the absorbance to increase to the pre-set value, e.g. 0.2 absorbance units. Figure 9.5 shows typical absorbance increases due to rising turbidity and to colour development during progression of the kinetic turbidimetric and colorimetric assays, respectively.

Kinetic chromogenic assays rely on software to undertake the reading of the plate, plotting of data and calculation of final test result, which is adjusted, where necessary, to account for any product dilution. The laboratory should ensure that all operators

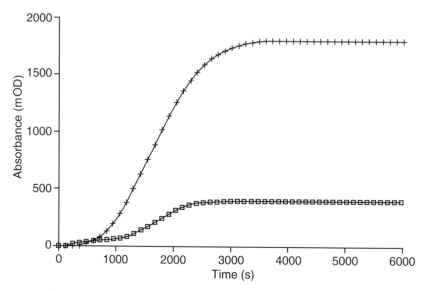

**Figure 9.5**  Absorbance increase during the progression of a kinetic turbidimetric assay ($\square$) and a kinetic chromogenic assay (+) in the presence of 0.5 EU ml$^{-1}$ endotoxin.

are fully conversant with the operation of the current software, and alternative packages should be compared prior to the purchase of a new system.

### 9.5.2  *Performance characteristics*

With respect to linearity and reproducibility, the performance criteria for the kinetic chromogenic assay are similar to those for the kinetic turbidimetric assay described in section 9.4.3. In order to verify the lack of product inhibition, an aliquot of test sample (or a dilution thereof) should be spiked with a known amount of endotoxin; a concentration of 0.5 EU ml$^{-1}$ is recommended. For samples which may contain a background endotoxin level $\geq 1$ EU ml$^{-1}$, the endotoxin spike should result in a final endotoxin concentration of 5.0 EU ml$^{-1}$. As with the kinetic turbidimetric assay, the endotoxin recovered is required by FDA criteria to equal the known concentration of the spike $\pm 50\%$, although the EP (1998) guideline allows a wider remit of 50–200% recovery.

Since the initial absorbance reading of each well is used as its own blank, samples which themselves possess significant colour do not present a special problem. If the background colour is 1.5 absorbance units, the sample should be diluted and re-assayed.

### 9.6  **Method selection**

Endotoxin testing began as a straightforward limit test, but it has become more complicated due to: development of several test methods; multiple reference standards in use; and increased scope of testing. Selection of the method should be determined by considerations of sample volume and type (water, final product, in-process,

**Table 9.7** Relative merits of *Limulus* amoebocyte lysate (LAL) testing methods

| Method | Advantages | Disadvantages |
|---|---|---|
| Gel-clot | Easy to interpret (yes or no result) Inexpensive reagents Low equipment cost Accepted by major pharmacopoeias | Subjective Semi-quantitative only at higher cost Product inhibition Limited throughput Cannot be automated |
| End-point chromogenic | Fast Quantitative Less product inhibition than gel-clot Objective Flexible equipment requirements | Higher cost than gel-clot Relatively small window of quantitation Most demanding method in terms of operator technique |
| Kinetic turbidimetric | Quantitative Can be automated User-friendly Inexpensive reagents Objective | Product inhibition (similar to gel-clot) Expensive equipment Relatively poor robustness |
| Kinetic chromogenic | Quantitative over large range Less product inhibition than gel-clot and kinetic turbidimetric methods User-friendly May be automated Robust | Relatively expensive reagents Expensive equipment |

devices) and of available equipment and budget. The relative merits of the four testing methods are summarized in Table 9.7.

### 9.6.1 *Water samples*

Water is involved in the manufacture of most, if not all, sterile products and it often represents the major proportion, up to 97%, of the final product. As a result, water testing often forms between 40 and 60% of all testing, and so must be as cost-effective as possible. It is necessary to have an effective monitoring system for feed water; breakdowns in water treatment systems frequently correlate with supply problems, especially following pipe works or water shortages. Most still manufacturers will only guarantee a 2 to 3 log reduction in bioburden, so if incoming feed water pyroburdens are around $250-300\,EU\,ml^{-1}$, it will prove difficult to achieve $0.25\,EU\,ml^{-1}$ for Water for Injection. Routine water testing results performed both by the gel-clot and the kinetic chromogenic method are shown in Figure 9.6; it is apparent that the greater sensitivity and quantitative nature of the kinetic assay affords advantages in the early recognition of a problem.

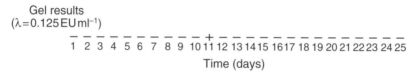

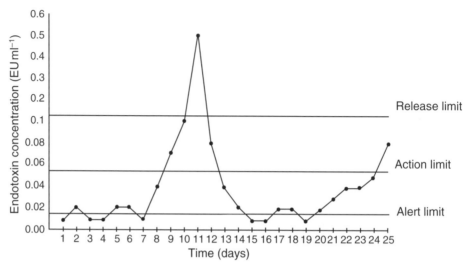

**Figure 9.6**   Routine endotoxin testing results from water. Note that the graph showing kinetic chromogenic data is much more informative than the gel results at the top.

Validation of water testing programmes may be undertaken by spiking the reservoir and demonstrating accurate and reproducible recoveries. Concentrated standardized endotoxin solutions containing $2000-7000 \, \text{EU ml}^{-1}$ are available commercially for this purpose. Spike recoveries resulting from automated kinetic assays may cause problems due to 'bowing' of the calibration plot (Figure 9.4 and section 9.4.1). The suitability of assay procedures for water and other samples is summarized in Table 9.8.

### 9.6.2   *Samples other than water*

Problems in endotoxin testing more commonly arise with raw materials than with water due to: their physical attributes, e.g. high viscosity, lack of solubility or extreme of pH; the adoption of unrealistically low limits which stretch detection levels; and difficulties in obtaining representative samples from bulk materials, especially with powders. The most appropriate testing method will depend partly on the nature of the raw material, but kinetic methods generally are found to be easier than end-point assays. Problems of interference may be overcome by dilution, although other solutions to the problem may also be required.

In-process samples are often the most variable, and some process steps, particularly for 'biological preparations', may involve high endotoxin levels which pose particular problems if high levels of protein are also present. A further complication is that there is often a need for quick results in order to avoid production delays. The gel-clot

**Table 9.8** Assay characteristics and method selection

| | Gel-clot | End-point chromogenic | Kinetic turbidimetric | Kinetic chromogenic |
|---|---|---|---|---|
| *Assay characteristics common to all sample types* | | | | |
| Detection level or range | $0.03$ EU ml$^{-1}$ fixed | $0.01 - 1.00$ EU ml$^{-1}$ | $0.01 - 10$ EU ml$^{-1}$ | $0.005 - 50$ EU ml$^{-1}$ |
| Operator involvement | High/very high | High | Medium | Medium |
| Ease of use | Low/very low | Low | High | High |
| Robustness | Medium | Low | Medium | High |
| Equipment cost | Low | Low/medium | High | High |
| | | | | |
| *Assay characteristics specific to water samples* | | | | |
| Problem areas | No indication of trends | Needs careful timing | Over-prediction of spike recovery | Over-prediction of spike recovery |
| Cost effectiveness | Variable | Medium | High | Medium to low |
| Overall rating* | | 3 | 1 | 2 |
| | | | | |
| *Assay characteristics specific to raw materials* | | | | |
| Problem areas | 1. Biologicals 2. Sensitivity | Short measuring range | 1. Biologicals 2. Interference | No special problems |
| Cost effectiveness | Variable | Low | Medium | High |
| Overall rating | 4 | 2 | 3 | 1 |
| | | | | |
| *Assay characteristics specific to in-process samples* | | | | |
| Problem areas | 1. Qualitative 2. Interference | Short measuring range | 1. Biologicals 2. Interference | No special problems |
| Cost effectiveness | Low | Medium | Low/medium | High |
| Overall rating | 4 | 3 | 2 | 1 |

* 1 = most suitable; 4 = least suitable.

method is rarely found to be suitable for in-process samples due its relative slowness and expense; the kinetic chromogenic assay is usually the preferred choice.

The difficulties posed by medical devices are often due to their large size or physical nature, e.g. gelatinous wound-healing agents or synthetic skin substitutes; these factors may render representative sampling the only realistic option. Spiking procedures and recovery studies on large surfaces are particularly difficult, and other minor problems are also encountered which are specific to device testing, e.g. locating a source of swabs which have acceptably low endotoxin levels.

## 9.7 Depyrogenation

This process originally centred around the use of dry heat ovens for vials and containers. More recently, however, it has become necessary to validate the use of depyrogenation tunnels and processes for determination of endotoxin on bungs, stoppers and other closures. Oven validations may be undertaken by preparing challenge vials coated with endotoxin. Generally stock solutions of approximately $10\,000\,\mathrm{EU\,ml}^{-1}$ are required; commercially available concentrated solutions are often found to save a considerable amount of time and effort. Depyrogenation tunnels and tunnel sterilizers are becoming increasingly used and they require careful validation due to the short transit times involved. Particular problems also exist with parenteral vial closures because of the materials used, e.g. natural rubber or polybutadiene are extremely variable. It is difficult to develop routine spiking and recovery methods for these materials, and high spiking values may not accurately determine the behaviour of closures in practice, especially if the pharmaceutical has surfactant properties.

## 9.8 Summary

There are currently four main LAL test methods in use today. These are the gelation test, the end-point chromogenic test, the kinetic turbidimetric test, and the kinetic chromogenic test. Each method has its strengths and weaknesses, and the laboratory should examine their requirements carefully before selecting a method of choice. Laboratories mainly involved in water testing, with a small number of products, will find that the gel-clot test provides the most cost-effective solution; for larger numbers the kinetic turbidimetric test is most useful. Laboratories involved in the measurement of raw materials, biological samples or a wide variety of products will find that the end-point or kinetic chromogenic methods yield the best results and ultimately the most cost-effective solution.

## References

BANG, F. (1956) A bacterial disease of *Limulus polyphemus. Bulletin of the Johns Hopkins Hospital*, **98**, 325–37.
*British Pharmacopoeia* (1999) The Stationery Office, London.
COOPER, J.F., LEVIN, J. and WAGNER, H.N. (1970) New rapid *in vitro* test for pyrogen in short-lived radiopharmaceuticals. *Journal of Nuclear Medicine*, **11**, 310.

DABBAH, R., FERRY, E., Jr, GUNTHER, D.A., HAHN, R., MAZUR, P., NELLY, M., NICOLAS, P., PIERCE, J.S., SLADE, T., WATSON, S., WEARY, M. and SANDFORD, R.L. (1980) Pyrogenicity of *E. coli* 055:B5 endotoxin by the USP rabbit pyrogen test – a HIMA collaborative study. *Journal of the Parenteral Drug Association*, **34**, 212–16.

*European Pharmacopoeia* (1997) 3rd edn. Strasbourg, Council of Europe, pp. 286–7.

GUILFOYLE, D.E. and MUNSON, T.E. (1982) Procedures for improving detection of endotoxin in products found incompatible for direct analysis with *Limulus* amoebocyte lysate. In: WATSON, S.W., LEVIN, J. and NOVITSKY, T.J. (eds), *Endotoxins and Their Detection With the Limulus Lysate Test*. Alan R. Liss, New York, pp. 79–88.

HORT, E. and PENFOLD, W.J. (1912a) Microorganisms and their relation to fever. *Journal of Hygiene*, **12**, 361–90.

HORT, E. and PENFOLD, W.J. (1912b) A critical study of experimental fever. *Proceedings of the Royal Society. London [Biology]*, **85**, 174–86.

HORT, E. and PENFOLD, W.J. (1912c) The relation of salvarsan fever to other forms of injection fever. *Proceedings of the Royal Society of Medicine* (Pt. 111), **5**, 131–9.

*Japanese Pharmacopoeia* (1996) 13th edn. Tokyo, The Society of Japanese Pharmacopoeia.

LEVIN, J. and BANG, F.B. (1964) The role of endotoxin in the extracellular coagulation of *Limulus* blood. *Bulletin of the Johns Hopkins Hospital*, **115**, 265–74.

LEVIN, J. and BANG, F.B., (1968) Clottable protein in *Limulus*; its localisation and kinetics of its coagulation by endotoxin. *Thrombosis, Diathesis and Haemorrhage*, **19**, 186–97.

Parenteral Drug Association (1980) A response to Federal Register 45, January 18, 1980. Interindustry communication.

SOLUM, N.O. (1970) Some characteristics of the clottable protein of *Limulus polyphemus* blood cells. *Thrombosis, Diathesis and Haemorrhage*, **23**, 170.

SOLUM, N.O. (1973) The coagulation of *Limulus polyphemus* hemocytes. A comparison of the clotted and non-clotted forms of the molecule. *Thrombosis Research*, **2**, 55.

TELLER, J.D. and KELLY, K.M. (1979) A turbidimetric *Limulus* amebocyte assay for the quantitative determination of Gram-negative endotoxin. In: COHEN, E. (ed.), *Biomedical Applications of the Horseshoe Crab (Limulidae)*. Alan R. Liss, New York, pp. 423–33.

United States Food and Drug Administration (1973) *Limulus* amoebocyte lysate: additional standards. *Federal Register*, **38**, 26130.

United States Food and Drug Administration (1977) Licensing of *Limulus* amoebocyte lysate: use as an alternative for the rabbit pyrogen test. *Federal Register*, **42**, 57789.

United States Food and Drug Administration (1987) A guideline on validation of the *Limulus* amoebocyte lysate test as an end-product endotoxin test for human and animal parenteral drugs, biological products and medical devices. Food and Drug Administration, Rockville, MD.

*United States Pharmacopoeia* (1995) 23rd edn. ⟨85⟩ Bacterial Endotoxins Test. US Pharmacopoeial Convention, Rockville, pp. 1696–7.

WEARY, M.E. (1996) Pyrogens and pyrogen testing. In: SWARBRICK, J. and BOYLAN, J.C. (eds), *Encyclopedia of Pharmaceutical Technology*, Vol. 13. Marcel Dekker, New York, pp. 179–205.

YOUNG, N.S., LEVIN, J. and PRENDERGAST, R.A. (1972) An invertebrate coagulation system activated by endotoxin: evidence for enzymatic mechanism. *Journal of Clinical Investigation*, **51**, 1790.

# 10

# Antimicrobial Preservative Efficacy Testing

NORMAN HODGES AND GEOFFREY HANLON

*School of Pharmacy and Biomolecular Sciences, Brighton University, BN2 4GJ, United Kingdom*

## 10.1   Introduction

A wide variety of products need to be protected from attack by micro-organisms during their period of use. This is both to protect the user from the dangers of infection and to prevent spoilage and deterioration of the product. In the case of medicines, foods and cosmetics, the safety of the user is the main priority, but maintenance of product quality and appearance and suitability of the product for its intended purpose are also important.

Preservatives are intended to protect the product from spoilage due to organisms introduced by the user and those which unavoidably arise during the manufacturing process; preservatives should never be used to counter poor manufacturing procedures or poor-quality ingredients. Clearly, sterile products in single dose units do not require preservation, neither do non-sterile single dose units such as tablets and capsules which are unlikely to sustain microbial survival provided that they are contained within suitable packaging. The need for a preservative system, therefore, most commonly arises if the product is to be subject to microbial challenge during repeated use. Some products are self-preserving, either because the active ingredients are inhibitory, the pH is inimical to growth, or because they contain high concentrations of sugar or other solutes which act as osmotic preservatives. These types of formulations are rare in the pharmaceutical arena, and the majority of multi-dose water-containing medicines incorporate chemical preservatives to prevent microbial spoilage.

The term preservative describes the *function* of a chemical agent in protecting a product from degradation or change which might arise if micro-organisms were to gain access and grow in it. However, this can be misleading since it might be thought that preservatives merely maintain the *status quo* (prevent micro-organisms growing, but not necessarily kill them), and as a result it is not uncommon to encounter the phrase *preservative levels of biocide* implying low concentrations of chemical agents which have only a bacteriostatic effect. In the majority of cases, however, the concentrations of preservatives used in product formulations are designed to give a rapid kill of any invading micro-organisms. Increasingly, preservatives are used in combination, with the intention of obtaining benefits such as a broader spectrum of antimicrobial cover or enhanced activity (synergy) and, occasionally, other chemicals which themselves have little or no antimicrobial activity are combined with a preservative in order to enhance its activity, e.g. EDTA (ethylene diamine tetra-acetic acid). Since more than one chemical is used to achieve the overall preservative effect, the term *preservative system* is sometimes employed.

There is an extensive range of chemical biocides available for potential use in preservative systems, but the particular requirements of preservatives within pharmaceutical formulations – and especially the need for lack of toxicity – are such that only a limited number of compounds are suitable. Some of the more common agents employed in the major product categories are shown in Table 10.1; they are listed in the order of frequency of use in the *United States Pharmacopoeia* (USP, 1995) formulae as revealed in a survey by Dabbah *et al.* (1996).

**Table 10.1**   Examples of preservatives commonly used in pharmaceutical formulations

| Product type | Preservative | Concentration (%, w/v) | Proportion of USP formulations in which preservative used (%)* |
|---|---|---|---|
| Parenteral | Benzyl alcohol | 0.1–3.0 | 31.0 |
| | Methyl/propyl paraben | 0.08–0.1/ 0.001–0.023 | 13.8 |
| | Phenol | 0.2–0.5 | 7.9 |
| | Methyl paraben (alone) | 0.1 | 6.6 |
| | Chlorbutanol | 0.25–0.5 | 5.3 |
| | Sodium metabisulphite | 0.025–0.66 | 5.3 |
| Ophthalmic | Benzalkonium chloride | 0.0025–0.0133 | 50.0 |
| | Thiomersal | 0.001–0.5 | 19.8 |
| | Methyl/propyl paraben | 0.05/0.01 | 6.6 |
| | Benzalkonium chloride plus EDTA | 0.01/0.1 | 3.3 |
| Creams | Benzyl alcohol | 1.0–2.0 | 25.4 |
| | Methyl/propyl paraben | NA | 18.6 |
| | Methyl paraben (alone) | 0.1–0.3 | 11.9 |
| | Benzoic acid | 0.2 | 8.5 |
| | Sorbic acid | 0.1 | 8.5 |
| | Chlorocresol | 0.05 | 6.8 |
| Oral | Sodium benzoate | NA | 34.4 |
| | Methyl/propyl paraben | NA | 18.3 |
| | Methyl paraben (alone) | 0.1 | 9.7 |
| | Methyl paraben plus sodium benzoate | NA | 7.5 |

*Only the most commonly used agents are listed; thus percentages in each product category do not total 100%.
NA = not available.

### 10.1.1 *The requirement for a biological assessment of preservative activity*

During product development it is necessary to demonstrate that the preservative system included within a formulation remains effective over prolonged periods of storage. Measurement of the concentration of a preservative within a product is a relatively simple procedure, and accurate chemical assays are available for most of the agents commonly used. However, this information is of limited value since the biocidal activity of the compound may be markedly influenced by the physical characteristics and the other ingredients within the formulation and the materials which comprise the packaging of the product. Factors which will influence activity include the following:

- *pH*: some preservatives such as benzoic acid and sorbic acid are profoundly affected by pH because this influences the ionic state of the compound. In order to enter a microbial cell and exert an antimicrobial effect, weak organic acids must be unionized, and this only occurs under conditions of low pH. At high pH values the biocidal activity of these weak acids is much reduced. It is possible, therefore, that during storage the pH may increase but the concentration of the preservative remains unchanged. The consequent reduction in biological performance would not be detected by chemical assay.

- *Presence of non-aqueous phase:* creams and emulsions have non-aqueous phases which do not usually require preserving since micro-organisms will grow almost exclusively in the aqueous phase of the formulation or at the interface. However, depending upon the partition coefficient of the preservative, some of the biocidal action within the aqueous phase may be lost if the compound is highly soluble in the oil phase. Partition can be an extended process encouraged by certain storage conditions.

- *Adsorption to suspended solids:* suspended solids contained within suspensions may adsorb preservatives, thus reducing their concentration in the aqueous phase to subeffective levels.

- *Adsorption onto plastic packaging:* glass containers do not usually represent a problem so far as the formulation is concerned, but with the trend towards increasing use of plastics, consideration must be given to the choice of preservative. Some compounds may be absorbed into the plastics and subsequently evaporate out from the container. Others may simply adsorb onto the inside of the container and be lost from solution.

In each of the cases described above the problem is exacerbated if the preservative is in low concentration or has a high concentration exponent (Denyer and Wallhaeusser, 2001). In the latter case a small decrease in concentration would dramatically reduce activity. It is evident, therefore, that because of the complexities of pharmaceutical preparations a simple chemical assay would not provide adequate information on the performance of a preservative system. Consequently, the only reliable method for assessing performance is a biological assay; these are described in the various pharmacopoeias and are commonly known as preservative efficacy tests or challenge tests. In principle, they are quite simple: standardized suspensions of bacteria, yeasts or moulds are inoculated into individual portions of the product to be tested and the survival of each organism is determined at intervals over a 28-day period. The product is considered to be adequately preserved (passes the test) if the levels of survivors are no greater than the values specified in the pharmacopoeia for that product type.

## 10.1.2 *Limitations of preservative efficacy tests*

Despite their apparent simplicity, preservative efficacy tests pose a number of practical problems and, of necessity, make certain assumptions. Ideally, one would expect that a preservative efficacy test would simulate the in-use situation for the product; this would entail, for example, repeated inoculation of mixed cultures at low concentrations. However, it may be difficult or impossible to distinguish the colonies of organisms used in a mixed culture so their relative survival capabilities could not easily be quantified, and low inocula would not readily permit a demonstration of the *extensive* microbial killing which is required in parenteral and ophthalmic products. The design of preservative efficacy tests, therefore, tends to be a compromise between realism and such practical considerations (Baird, 1995; Hodges and Denyer, 1996; Hodges, 1998).

Many preservatives in aqueous solution exhibit first-order kill kinetics. However, others do not, and so when log percentage survivors are plotted as a function of time (to give what is described as a kill curve or survivor plot), the graph may show the presence of shoulders (lag) or tails rather than the expected linear relationship. There are a number of reasons for these non-linear plots, including the presence of resistant cells within a population and the possibility that lysis of dead cells releases materials which afford protection to those still alive. Non-linear kill curves can make interpretation of preservative efficacy data complicated, particularly when only one or two time points are used for sampling. Figure 10.1, for example, illustrates the effect of a potassium sorbate/EDTA preservative system on the survival of *Candida albicans* during the preservative testing of a nasal product. Samples taken within the first 10 days would indicate that the preservative was almost totally ineffective, but that would not be the conclusion if the product was sampled at day 14 and subsequently.

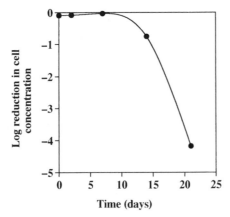

**Figure 10.1** The survival of *Candida albicans* in a potassium sorbate-preserved nasal product.

Preservative efficacy tests also make certain assumptions about the surviving population of cells. For instance, they assume that the survivors have similar growth characteristics to the inoculum culture. However, those cells which are still viable after 28 days in contact with the product will have been damaged by the preservative and may require different recovery conditions from the original population in order to grow and form visible colonies (Baird, 1995). Such phenomena may enhance preservative efficacy test data where inappropriate recovery media are used and there is an expectation of growth within 48 h. Falsely optimistic data may also result from those situations where a preservative system causes the cells to clump together. As the methods for assessing survivors are usually plating procedures, such aggregation will give artificially low counts. Alternatively, aggregation may contribute to non-linear kill curves because the larger aggregates protect those cells within the centre and lead to apparently high levels of resistance – at least during the initial period of exposure.

Concerns have also to be expressed about the nature of the inoculum cultures, which are typically healthy, vigorous micro-organisms grown rapidly under favourable conditions. In reality, the micro-organisms entering a product during use are likely to be poorly nourished, slow-growing and possibly desiccated (Gilbert and Brown, 1995a). Since many preservatives are less active against slowly growing cells than fast-growing ones, authentic product contaminants may present a far more severe test of the preservative system than the cells actually used (see section 10.3.1). Not only is the growth rate a determinant of preservative sensitivity, but so too is the nature of the culture medium on which the cells are grown. Al-Hiti and Gilbert (1980) grew challenge test micro-organisms under conditions of carbon, nitrogen and phosphate depletion and found a marked difference in sensitivity. This highlights the need to define more adequately the growth media and conditions for the production of inocula. Products which are only marginally successful in passing a preservative test may fail when confronted with the sterner challenge of micro-organisms in their natural state.

During use, micro-organisms may also colonize the containers in the form of biofilms. This may be a particular problem with eye drops for example, where the dropper can easily make contact with the eye during use. Micro-organisms can adhere to the plastic nozzle and form resistant biofilms on the inner surface of the container. Recent studies have shown that approximately 10% of preserved eye drops contain micro-organisms when examined during use, and many of these were associated with the nozzle (Livingstone *et al.*, 1998).

For these and other reasons it is not possible for a preservative efficacy test to integrate and simulate all the parameters which may arise during actual usage, and the main purpose of the test is to provide, during the development stage of a new product, a relatively simple and reliable indication of the adequacy of protection against microbial spoilage which should be capable of replication to give the same results from one laboratory to another.

The first preservative efficacy test was described in the *18th United States Pharmacopoeia XVIII* in 1970, and was followed by those in the British, Italian and German Pharmacopoeias in the 1980s. Since these original tests, the procedures have been refined and attempts made to bring about harmonization. Differences between the various national European pharmacopoeias delayed the publication of *the* European Pharmacopoeial test until 1992. The USP test remained virtually unchanged for 25 years after its introduction, but the recent Eighth supplement to 23rd USP (1998) included a much revised test. Despite trends towards international harmonization, minor differences between the *European Pharmacopoeia* (EP, 1997) and the USP (1998) still exist in terms of methodology; these will be discussed later. The major differences lie in the criteria for passing the test (see section 10.2.8).

## 10.2   Test procedure and factors influencing reproducibility

Although preservative efficacy tests are simple in principle and the pharmacopoeial test protocols appear to be relatively precise, it is not at all uncommon for substantial differences in survivor levels to be recorded, even when the same product is examined on successive days in the same laboratory (Davison, 1996; Hodges *et al.*, 1996). Thus, the potential for variation *between* laboratories is even greater, and this may pose a particular problem where a product is tested in two or more laboratories within a large company – especially if they are on different sites or in different countries. Consequently, it is essential that:

- persons conducting a preservative efficacy test are aware of the features of the test which can lead to day-to-day variability in results; and

- company microbiologists adopt testing protocols which, while conforming to the requirements of the relevant pharmacopoeia, are even more precisely defined with respect to the major factors influencing reproducibility.

Initial inspection of the test procedures described in a single pharmacopoeia might create the impression that the experimental details are well defined, but closer examination reveals a choice of viable counting methods and preservative neutralization procedures together with relatively wide permissible ranges for inoculum concentrations, culture incubation conditions and product storage. All these factors, and others not mentioned at all in official protocols, may be a source of variation.

### 10.2.1   *Selection of viable counting method and demonstration of operator competence*

The principal microbiological technique involved in preservative testing is viable counting; thus the selection of the most appropriate method and demonstration of operator competence are of crucial importance. However, in many laboratories, alternative viable counting methods are not considered, and the method 'selected' is, by default, that which has always been used, or the only one familiar to the personnel – even though it may not be the most appropriate. Furthermore, there is no pharmacopoeial requirement to verify that the proposed operator is proficient at viable counting. Clearly, if an operator cannot achieve satisfactory precision when performing replicate counts on a standard bacterial suspension, they are unlikely to achieve good reproducibility when undertaking the more complex manipulations in preservative efficacy tests where, for example, such factors as product viscosity may predispose to inaccurate sample volumes.

   There are three possible viable counting procedures which can be considered for an EP test (pour plates, surface spread plates and membrane filtration methods), but the USP test directs that the 'plate count' method be used, and this, as described in ⟨61⟩ *Microbial Limit Tests*, means pour plates not surface spread plates. It should be emphasized that these alternatives do not necessarily give the same result when applied to the same sample (Cowen and Steiger, 1976). Pour plate methods are in common use and afford the advantages of simplicity and the ability to detect lower concentrations than surface spread methods. Their disadvantage is that the colonies of strictly aerobic organisms (*Pseudomonas aeruginosa* for example) may be very small and consequently overlooked if they arise in the deepest part of the agar. Surface spread plates tend to give larger colonies which are easier to count and which often possess visible distinguishing features (e.g. shape, pigmentation, surface markings), so chance contaminants can be more readily distinguished. Membrane filtration methods can detect lower concentrations of organisms than surface spread or pour plates and are the best means of eliminating antimicrobial activity in the sample because the organisms are physically separated from the preservative (see neutralization of preservative activity, section 10.2.5). However, the 47-mm diameter membranes, which are most commonly used, afford a surface area which is only 27% of that on a standard Petri dish, so crowding of colonies is more of a problem, particularly with moulds such as *Aspergillus niger*, where the colonies may become confluent if incubated for too long.

What constitutes adequate precision in viable counting is, of course, debatable, and there are no clear guidelines in official compendia. It is not at all difficult for a competent operator to conduct six or more replicate viable counts on a bacterial suspension at a concentration of approximately $10^8 \, ml^{-1}$ and record a standard deviation which is less than 10% of the mean (coefficient of variation), and this might be regarded as a simple criterion of competence. Coefficients of variation of less than 10% can be more difficult to achieve with moulds where individual colonies may be difficult to distinguish, and because of the larger colony size, plates with fewer colonies are selected for counting.

The Tenth Supplement to the 23rd edition of the USP (1999) describes a validation procedure which is intended to demonstrate that a preservative has been adequately neutralized and that a chemical inactivator used for the purpose, is itself, non-toxic. This procedure states that adequate neutralization is achieved if the recovery of a low inoculum [10–100 colony forming units (c.f.u.)] exposed to the neutralization procedure is not less than 70% of the (unexposed) control. This implies that counts which should really be the same should not differ from each other by more than 30%, so this, too, may be taken as a guide figure by which to assess operator technique.

### 10.2.2  *Selection and maintenance of test organisms*

One of the desirable characteristics of a preservative efficacy test is that it should produce the same result when applied to the same product sample tested by different laboratories. In order to achieve this, it is essential that the same test organisms are used in each laboratory and that the resistance and growth properties of the organisms are standardized and stable. Consequently, the trends in pharmacopoeial testing protocols are towards the use of test cultures which are only a limited number of passages (subcultures) removed from the source (obtained from an internationally accessible culture collection) and the removal of the requirement/recommendation to employ strains that might arise as product contaminants during manufacture and use.

The organisms to be used for testing are the same in the European, Japanese and United States Pharmacopoeias, and the strains specified may be obtained from the

**Table 10.2**  Preservative testing strains specified in the EP and USP

| Organism | EP (1997) | USP (1998) |
|---|---|---|
| *Pseudomonas aeruginosa*<br>ATCC 9027; NCIMB 8626: CIP 82.118 | All products | All products |
| *Staphylococcus aureus*<br>ATCC 6538; NCTC 10788; NCIMB 9518:<br>CIP 4.83 | All products | All products |
| *Candida albicans*<br>ATCC 10231; NCPF 3179: IP 48.72 | All products | All products |
| *Aspergillus niger*<br>ATCC 16404; IMI 149007; IP 1431.83 | All products | All products |
| *Escherichia coli*<br>ATCC 8739; NCIMB 8545: CIP 53.126 | Oral products | All products |
| *Zygosaccharomyces rouxii*<br>NCYC 381: IP 2021.92 | Oral products with<br>high sugar content | Not used |

| | |
|---|---|
| ATCC | American Type Culture Collection (USA). |
| NCIMB | National Collection of Industrial and Marine Bacteria (UK). |
| NCTC | National Collection of Type Cultures (UK). |
| CIP and IP | Collection of the Institut Pasteur (France). |
| NCPF | National Collection of Pathogenic Fungi (UK). |
| IMI | Imperial Mycological Institute (now CABI Bioscience, UK). |
| NCYC | National Collection of Yeast Cultures (UK). |

sources indicated above (Table 10.2). Individual rather than mixed inocula are used for the reasons described in section 10.1.2; organisms are inoculated into separate containers of the product under test.

A strict interpretation of the USP test would require the cultures to be obtained from the American Type Culture Collection (ATCC) alone, since there is no indication in the test protocol that equivalent strains from other collections are acceptable. The current edition of the EP (1997) (but now not the USP) suggests inclusion of additional organisms representing the contaminants which might arise during manufacture, storage or use. The disadvantages of this are considered in section 10.3.

It has long been recognized that repeated subculture of any particular strain of micro-organism may lead to the selection of cells which differ quite significantly from the starting culture in terms of growth characteristics and resistance to preservatives. In the EP (1997) there are no precise requirements concerning the number of subcultures from which the test organisms may be removed from the original, merely a direction that subcultures should be 'kept to a minimum'. In contrast, the USP (1998) requires that all organisms used in the test must not be more than five passages (subcultures) from the original ATCC culture. Compliance with such a requirement would not necessitate unacceptably frequent repurchase of cultures from the culture collection if a well-organized maintenance procedure was used; indeed, the USP (1998) gives details of seed lot techniques for culture cryopreservation with glycerol at $\leq 50°C$. All test organisms would be expected to survive for at least 6 months if stored in well-sealed slopes at 4°C, so it would be possible to maintain 'master' slopes in this way, from which 'working' slopes could be prepared on a monthly or more frequent basis. The purity of these cultures would need to be confirmed by visual inspection of the colonies arising on streaked plates.

### 10.2.3  *Growth, standardization and storage of test inocula*

The EP (1997) and USP (1998) differ in several details regarding the preparation of inocula; these are identified in Table 10.3.

The same media are recommended in both pharmacopoeias for the preparation of the test inocula, although the USP specifies the need to demonstrate growth-promoting properties. The USP option of using liquid media rather than agar might also lead to differences in preservative susceptibility of the inocula due to variations in growth rate (Al-Hiti and Gilbert, 1983). Furthermore, the use of liquid cultures necessarily involves a centrifugation step, and the effect that this may have on the viable count should not be overlooked; Gilbert *et al.* (1991) showed that centrifugation could reduce the viable count of such common bacteria as *Pseudomonas aeruginosa*. The USP does not specify whether liquid cultures should be incubated static or shaken, so oxygen availability might limit culture growth in a static culture, particularly in the case of *Pseudomonas aeruginosa*, which is a strict aerobe.

**Table 10.3**  Procedures for the preparation and storage of inocula

|  | EP (1997) | USP (1998) |
|---|---|---|
| Recommended bacterial medium | Soybean casein digest agar (tryptone soya agar) | Soybean casein digest agar (tryptone soya agar) |
| Recommended yeast and mould medium | Sabouraud dextrose agar without antibiotics | Sabouraud dextrose agar |
| Solid or liquid media | Only solid media | Liquid media permissible |
| Inoculum culture incubation conditions | Bacteria – 30–35°C for 18–24 h | Bacteria – 30–35°C for 18–24 h |
|  | *C. albicans* – 20–25°C for 48 h | *C. albicans* – 20–25°C for 44–52 h |
|  | *A. niger* – 20–25°C for one week or until good sporulation | *A. niger* – 20–25°C for 6–10 days |
| Medium for harvesting and suspending inocula | Peptone-saline for bacteria and *C. albicans*; saline plus 0.05% polysorbate 80 for *A. niger* | Saline alone for bacteria and *C. albicans*; saline plus 0.05% polysorbate 80 for *A. niger* |
| Standardization and use | Adjust inoculum suspension to approximately $10^8$ c.f.u. ml$^{-1}$ | Adjust inoculum suspension to approximately $1 \times 10^8$ c.f.u. ml$^{-1}$; note the difference in precision |
| Storage | Use immediately | If not used within 2 h, bacteria and yeast suspensions may be stored in a refrigerator for 24 h and *A. niger* for up to 7 days |

The incubation conditions recommended by the two pharmacopoeias are the same (except for the extended time permitted by the EP to achieve good *Aspergillus niger* sporulation), although the allowed temperature and time ranges are sufficiently wide to permit marked differences in the growth rates and physiological states of the cultures when they are harvested. It is quite feasible that bacterial cultures grown on agar at the minimum temperature (30°C) for the minimum time (18 h) might contain a high proportion of cells still in active growth, but incubation using the maximum temperature (35°C) and time (24 h) is much more likely to result in cell growth being complete. Since it is well established that the susceptibility of micro-organisms to antimicrobial chemicals may be markedly affected by growth rate (Gilbert and Brown, 1995b), it is desirable to standardize, as completely as possible, the actual temperatures and incubation times employed.

While it is possible to design a test protocol which simultaneously conforms in every other respect to both the EP and the USP, it is not possible to reconcile the different suspending media recommended in the two pharmacopoeias. The inclusion of peptone in the medium specified in the EP method for bacteria and *Candida albicans* means there is the possibility of significant cell growth after harvesting if the suspensions of test organisms are kept for several hours in a warm laboratory prior to use. As a consequence, the EP directs that the inocula should be used 'immediately', whereas the USP accepts storage of inocula since growth in saline without peptone is unlikely. It should also be recognized that storage of inocula leads to starvation and changes in microbial physiology.

Standardizing the suspensions at a concentration of approximately $10^8$ c.f.u. ml$^{-1}$ is readily achieved for bacteria by use of a calibration plot relating viable count to turbidity. Figure 10.2 shows typical calibration plots for the three bacterial strains commonly used for testing. Clearly, the strain having the smallest volume and surface area would be expected to generate the lowest turbidity at any given concentration of c.f.u. ml$^{-1}$. On this basis, the data suggest that *Staphylococcus aureus* is the organism possessing the largest cells. However, this may be a reflection of the fact that its cells have a marked tendency to aggregate, which results in a large surface area for the c.f.u. Once the preparation of the inoculum suspension has been performed several times,

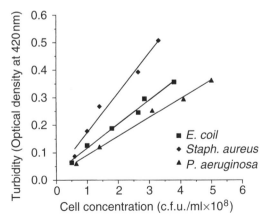

**Figure 10.2** Specimen turbidity calibration plots for the bacteria used in the EP and USP preservative efficacy tests.

it is possible to prepare a standard operating procedure for harvesting and dilution of the cultures which is sufficiently reliable to make a calibration plot unnecessary. Nevertheless, spectrophotometric measurement of optical density of the inoculum suspension provides a simple and quick means of standardizing and checking the microbial challenge, and this is now specifically mentioned in the USP (1998) (see also Chapter 4).

Standardizing suspensions of yeasts and *Aspergillus niger* spores is also possible using turbidity, but the extent to which yeast cells exhibit budding and mould spores aggregate (clump) is so variable that the most precise standardization is achieved by direct counting of the individual cells or c.f.u. using a haemocytometer counting chamber. Another reason to avoid reliance on turbidity for standardizing mould spore suspensions is that fragments of the mycelium may contribute to the observed turbidity and lead to an over-estimate of the spore concentration. Furthermore, mycelial fragments may be capable of forming colonies on the plate and so contribute to a high baseline or time zero count, but because vegetative mycelium is likely to be more preservative-sensitive than spores, the count would fall quickly during the early period of exposure.

After dilution of the suspension to a concentration of *approximately* $10^8$ c.f.u. ml$^{-1}$ the accurate figure is determined by performing a viable count on this suspension. This count is used as the reference value by which the degree of inactivation (killing) of the inoculum in the product is assessed; ideally, the count should be performed in a manner which reproduces, as far as possible, the dilution scheme and volumetric errors inherent in the counting of the inoculated samples.

### 10.2.4 *Test container, product inoculation, mixing and storage*

Preservative efficacy tests normally involve the inoculation of product in its final market container in order to allow any interaction between container and product which might modify preservative activity to take place (e.g. 'sorption of preservative onto or into the container or leaching of materials from the container which may interact with the preservative). Since plastics are increasingly common as packaging materials it is important to allow an opportunity for these interactions to occur, otherwise a false impression of the adequacy of preservation may be given. Thus, even when products cannot be tested in their final market pack – owing to difficulties in ensuring adequate mixing of the inoculum in ointments for example – it is essential that the preservative testing is performed on a product sample that has been kept in the market container for as long as possible (adsorption isotherms show that much of the preservative loss occurs in the first few weeks of storage).

Where the market container cannot be used, a substitute container must be selected and this should be of borosilicate glass (rather than soda glass which may release alkali during storage). Wide-mouthed, ointment jars which have smooth (rather than ribbed) sides are generally found to be the most convenient substitute containers since they will allow the use of stirring rods to permit mixing of the inoculum with viscous products. Adequacy of mixing can be assessed by addition of a non-toxic indicator, either incorporated into the inoculum or as a separate solution added to a replicate container of product: phenol red is suggested in the *Japanese Pharmacopoeia* (1996).

In order to achieve an initial cell concentration in the product of $10^5$–$10^6$ c.f.u. ml$^{-1}$, the normal procedure is to add a volume of inoculum suspension (adjusted to $10^7 - 10^8$ c.f.u. ml$^{-1}$) which represents 0.5–1.0% by volume of the product. In this respect it is important to confirm that the volume of product in the pack is exactly and reproducibly the value stated on the label. Frequently, preservative tests are performed on products in small pack sizes of the order of 10–20 ml, so the volume of inoculum to be added can be as small as 50 μl. This can be measured with a properly calibrated micropipette, but addition by weight is a good alternative since the weight per ml of the inoculum suspension can safely be assumed to be unity and the volume added does not have to be *exactly* 50 μl provided that it is accurately known.

Again, the relatively wide permissible concentration of $10^5 - 10^6$ c.f.u. ml$^{-1}$ provides scope for inconsistencies between replicate tests. Many textbooks indicate that microbial killing is likely to follow first-order kinetics; this means that the death *rate* is the same regardless of inoculum level. However, theory would predict that the interaction should follow second-order kinetics since the death rate will be influenced both by the concentration of preservative chemical and the concentration of molecules in the cell with which the preservative interacts. First-order kinetics will only arise if the preservative is present in gross excess and its concentration is effectively constant because so little of it is 'used up' when it interacts with the cell. Second-order kinetics means that there comes a point at which the inoculum concentration *does* influence the death rate, and raises the possibility that inoculum concentrations at the top of the permitted range, i.e. near to $10^6$ c.f.u. ml$^{-1}$, may result in a less impressive preservative performance than those nearer to $10^5$ c.f.u. ml$^{-1}$.

The USP preservative efficacy test is intended for application to anhydrous as well as water-based products. The value of indiscriminate testing of anhydrous products is questionable because most of such products are not vulnerable to microbial spoilage. The justifications offered for testing anhydrous products are: (i) that some anhydrous products contain hygroscopic materials; and (ii) while micro-organisms will not actually grow without water, they will not necessarily die either, so, in the absence of a preservative there is nothing to remove a high and possibly hazardous bioburden. Testing anhydrous products poses a problem with respect to the introduction of the inoculum, since the product obviously ceases to be strictly anhydrous if even 0.5% by volume of aqueous microbial suspension is added to it. It may be difficult, if not impossible, to achieve uniform mixing of an aqueous inoculum with materials such as ointment bases. Furthermore, the preservative must partition from the oily phase into the aqueous phase before it can act on the inoculum cells. However, problems also occur in the alternative strategy of preparing inoculum suspensions in non-aqueous solvents, since almost all common solvents possess some antimicrobial activity. Consequently, the addition of an aqueous inoculum is usually accepted as the least unsatisfactory method and the USP (1998) recommends the addition of a surfactant to the inoculum which does not affect survival of the micro-organism or potentiate the preservative. Heating solid ointment bases to $47.5 \pm 2.5°C$ is also a USP-recommended procedure for facilitating incorporation of the inoculum.

The storage temperature recommended for the inoculated product is 20–25°C; the EP directs that the stored product is protected from light, the USP does not. The temperature coefficients for several preservatives ($Q_{10}$ values) are sufficiently large that a 5°C temperature rise will increase the killing rate five-fold or more (Denyer and Wallhaeusser, 1990). Thus, it is quite feasible that two tests conducted on the same product at the maximum and minimum temperatures could give different results. Consequently, it is essential that the storage temperature is controlled as precisely as possible in order to achieve reproducibility of test data.

### 10.2.5 *Product sampling and preservative neutralization*

The EP test requires sampling of the inoculated product at time zero, i.e. as soon after inoculation as possible, and at two or three intervals thereafter (from 6 h, 24 h, 2 days, 7 days, 14 days, and 28 days) depending upon the type of product under test. The USP protocol describes no time zero sample, but there is a requirement that samples are taken at 7, 14 and 28 days after inoculation depending on product category. The value of the EP time zero sample should be questioned because:

- in reality, it can never be taken at time zero anyway;
- the time zero count is not used as the baseline value by which the extent of microbial killing is assessed; and
- its purpose is unclear.

The time zero sample is often taken at least 1 min after inoculation, and sometimes several minutes are required to ensure adequate mixing of the inoculum, especially with viscous creams. This delay may be sufficient to permit a significant degree of microbial killing, so that the time zero count falls far below the expected value calculated from the inoculum suspension count. Thus, it is often observed that one or more organisms (particularly *Pseudomonas aeruginosa* since the strain employed is relatively sensitive to most of the common preservatives) exhibits a time zero survival which is less than 50%, and sometimes very much lower values are encountered.

It is essential that the antimicrobial activity of the preservative is stopped at the time that the sample is taken. If the preservative continued to exert its activity during the dilution, plating and incubation of the sample then the recorded level of survivors would be artificially low, i.e. the preservative would appear to be more active than it really was. Stopping the continued preservative action is variously referred to as neutralization, inactivation or quenching of the preservative, and it can be achieved in one of three ways.

1 Dilution of the product sample so that the preservative is no longer at an effective concentration.
2 Chemical neutralization.
3 Membrane filtration of the (usually diluted) product sample so that the surviving organisms are retained and washed on the surface on the membrane filter and the preservative is thus physically separated from them.

Dilution is, of course, the easiest option, but it is appropriate only for those preservatives with a high concentration exponent (alcohols and phenolics, for example). For preservatives whose activity is little changed by dilution (low concentration exponent), dilution is not an option. If dilution is to be relied upon to eliminate preservative activity, it is important to validate the procedure for the situation which arises towards the end of the preservative test when the level of survivors is low. Here, a countable number of colonies may result from only a single 10-fold dilution of the sample, whereas two or three such decimal dilutions might have been required at the start; the dilution used at this later stage of the test may therefore be insufficient to achieve complete neutralization.

Chemical neutralization is commonly employed, and it is the only method of preservative inactivation mentioned in the current USP test which requires addition of a specific chemical inactivator (if there is one available) to the plate count medium or the diluent. The USP makes no mention of what to do if there is no such inactivator available. Lists of inactivators which may be employed to neutralize specific preservatives are to be found in several publications (Russell 1981; Denyer and Wallhaeusser, 2001; Baird, 1995; Hugo and Russell, 1998), although a mixture of lecithin and polysorbate 80 is so frequently found to be satisfactory that media containing this combination are commercially available (letheen broth and agar; Difco, West Molesey, Surrey, UK).

Membrane filtration is considered in the EP sterility test procedure to be the most effective means by which to avoid preservative carry-over. The cells retained on the surface of the membrane can be washed with a simple solution such as peptone or one containing a preservative neutralizer. There is no certainty, however, that the washing procedure will necessarily remove all the preservative that might have become adsorbed or covalently linked to polymers on the surface of the cell or the membrane itself. The USP (1998) does not mention membrane filtration as a means of eliminating antimicrobial activity in a preservative efficacy test.

### 10.2.6  Incubation conditions for organisms recovered from inoculated product

Just as there are permissible ranges of both temperature and time in pharmacopoeial protocols for growth of inocula, there are also ranges adopted for incubation of samples taken during the testing process. The recommended pharmacopoeial incubation conditions are shown in Table 10.4.

It is unlikely that the recorded plate count for the inoculum culture will vary to any great extent, regardless of the actual incubation temperature and time selected from the permissible ranges. However, it is well established that cells which have survived sublethal exposure to antimicrobial treatments may have incubation optima which differ from those of the unexposed cells (Baird, 1995); if the optima are changed, there is usually a requirement for a longer period of incubation (see, for example, pharmacopoeial sterility testing protocols). This may mean that while the inoculum suspension will give the maximum potential colony count when incubated for the shortest time at the minimum permitted temperature, the preservative-treated cells would give a submaximal count which could lead to the preservative performance being artificially high and, if different laboratories selected different incubation conditions, this may contribute to inter-laboratory variation.

**Table 10.4** Pharmacopoeial recommended incubation conditions for samples taken during testing

|  | EP (1997) | | USP (1998) | |
| --- | --- | --- | --- | --- |
|  | Bacteria | Yeast/moulds | Bacteria | Yeast/moulds |
| Temperature | 30–35°C | 20–25°C | 30–35°C | 20–25°C |
| Time | 5 days, unless a more reliable count is obtained in a shorter time | 5 days, unless a more reliable count is obtained in a shorter time | 3–5 days | 3–5 days for *C. albicans*; 3–7 days for *A. niger* |

### 10.2.7  *Validation*

For the test procedure to be considered valid it is necessary to show:

- that there is adequate mixing of the inoculum with the product (see section 10.2.4);
- the effectiveness of any specific inactivator used to neutralize the preservative;
- that the inactivator is not, itself, toxic to micro-organisms; and
- the ability of the procedure to demonstrate the required reduction in count of viable micro-organisms.

Although the current EP indicates a requirement for these controls, it does not specify precisely how they should be undertaken; consequently, this is an aspect of the test which poses problems, particularly to inexperienced operators. Currently the clearest advice on validation is that contained in the Tenth Supplement to the 23rd edition of the USP (1999) which compares the recovery of micro-organisms exposed to the neutralization procedure with an unexposed control. If the recovery of a low inoculum (10–100 c.f.u.) is not less than 70% of the expected value the neutralizer is considered to be effective and non-toxic to the test organism. The ability to recover such an inoculum is also taken to show the ability of the media to detect small numbers of viable organisms, and therefore the ability of the method to demonstrate the required reduction in count. At least three independent replicates of the experiment are required to demonstrate that the recoveries of the neutralizer-treated and untreated groups are similar; larger numbers of replicates render the results amenable to statistical analysis.

If data show that one of the pharmacopoeial test organisms is more sensitive to the preservative than the others, it may be tempting to conduct the validation using that organism alone – on the basis that if it gives satisfactory results the other organisms would too. However, there are two possible flaws in this approach:

1   The organism which is most sensitive to the preservative is not necessarily the one which is most sensitive to any inhibitory effects that the neutralizer may possess.
2   The rank order of sensitivity of the organisms to the preservative may be concentration-dependent. Consequently, it is not possible to use data obtained at one preservative concentration to predict which will be the most sensitive organism at another.

The majority of preservatives in common use kill the test bacteria far more quickly than they kill *Candida albicans* or *Aspergillus niger*, and so it is extremely unlikely that the last two named species would be the most sensitive to the preservative or to the neutralizer. Consequently, it may be considered reasonable to conduct the validation experiments on the bacteria alone rather than on all the test species. However, for the reasons described above, selecting a single bacterial species for validation purposes may prove to be a mistake.

### 10.2.8  *Interpretation of test results*

The EP (1997) describes two sets of acceptance criteria for parenteral/ophthalmic preparations and for topical preparations as shown in Table 10.5. The A criteria are more stringent and express the recommended efficacy to be achieved. However, in certain justified cases, such as where there may be a risk of adverse reactions, a less demanding set of conditions (B criteria) may be acceptable to the licensing authority. A single set of criteria are laid down for oral preparations. The USP (1998) performance criteria are, for each category of product, less demanding than those of the EP (1997) (Table 10.6), although it must be recognized that the USP criteria should be met by *every* batch of *every* USP product (and the same standards would be expected by the licensing authorities for non-pharmacopoeial preparations).

It is interesting to note that the EP does not offer a definition of the terms 'no recovery' and 'no increase', although the proposals described in Pharmeuropa (1994) tend to be used in practice (no recovery = $4\log_{10}$ cycle reduction or greater; no increase = not more than $0.3\log_{10}$ (approximately 100%) higher than the last value for which a criterion is given). The USP specifies 'no increase' as not more than $0.5\log_{10}$ higher than the previous value measured. There is a difference, too, in terms of the precision required for the $\log_{10}$ reductions specified in the two pharmacopoeias. The EP expresses log reductions to the nearest integer, i.e. 1, 2 or $3\log_{10}$ units, but the USP specifies the reductions to 0.1 of a unit, e.g. 1.0, 2.0 and 3.0. The USP notation would require for example, that if the calculated reduction in bacteria for a parenteral product at 14 days was 2.94 units, the product would fail because this would be rounded down to 2.9, but if it were 2.96 the product would pass since this would be rounded up to 3.0. This suggests a degree of precision in bacterial counting which is unlikely to be achieved, and is at odds with the definition of 'no increase' which suggests that two values which differ by $0.5\log_{10}$ units may be regarded as not significantly different.

The requirements of the *British Pharmacopoeia* (1999) are identical to those of the EP (1997), except that the BP includes performance criteria for ear preparations. This requirement should not be overlooked by manufacturers exporting to countries of the British Commonwealth, some of which retain the BP criteria for regulatory purposes.

**Table 10.5** EP (1997) and BP (1999) performance criteria for preservative efficacy

| Product type | Micro-organism | | Log reductions | | | | | |
|---|---|---|---|---|---|---|---|---|
| | | | 6 h | 24 h | 48 h | 7 days | 14 days | 28 days |
| Parenteral/ophthalmic | Bacteria | A | 2 | 3 | | | | NR |
| | | B | | 1 | | 3 | | NI |
| | Fungi | A | | | | 2 | | NI |
| | | B | | | | | 1 | NI |
| Topical | Bacteria | A | | | 2 | 3 | | NI |
| | | B | | | | | 3 | NI |
| | Fungi | A | | | | | 2 | NI |
| | | B | | | | | 1 | NI |
| Oral | Bacteria | | | | | | 3 | NI |
| | Fungi | | | | | | 1 | NI |
| Ear preparations | Bacteria | | 2 | 3 | | | | NR |
| BP (1999) only | Fungi | | | | | | 2 | |

NR = no recovery.
NI = no increase (see text).

**Table 10.6** USP (1998) performance criteria for preservative efficacy

| Product category | Organism | Log reductions | | |
|---|---|---|---|---|
| | | 7 days | 14 days | 28 days |
| Category 1A* | | | | |
| Injections, parenteral | Bacteria | 1.0 | 3.0 | NI |
| emulsions, otic, sterile nasal and ophthalmics | Fungi | NI | NI | NI |
| Category 1B* | | | | |
| Topicals, non-sterile | Bacteria | | 2.0 | NI |
| nasal products, emulsions (including those applied to mucous membranes) | Fungi | | NI | NI |
| Category 1C* | | | | |
| Oral products | Bacteria | | 1.0 | NI |
| | Fungi | | NI | NI |
| Category 2 | | | | |
| Non-aqueous | Bacteria | | NI | NI |
| products | Fungi | | NI | NI |

\* Products made with aqueous vehicles or bases.
NI = not more than 0.5 $\log_{10}$ unit higher than the previous value measured.


## 10.3    Adaptations and alternatives to pharmacopoeial tests

The preservative testing philosophies adopted in Europe and the USA differ in that the EP test is non-mandatory, i.e. there is not a requirement that every batch of every preserved pharmacopoeial product should pass the test; indeed, it is recognized that some pharmacopoeial formulations are unlikely to do so, e.g. insulin injection. Thus, the EP test is not used as part of the routine quality assurance programme associated with manufacture, but as part of the product development programme. For this reason there is no absolute requirement that the pharmacopoeial protocol be rigidly adhered to, and the option exists for the pharmacopoeial test to be adapted or extended yet remain acceptable for product registration purposes. This feature of the test was more in evidence in the earlier BP protocols of the 1980s; these included recommendations to consider the use not only of additional test organisms, but extended sampling periods (up to 3 months) and the repeated inoculation of the same sample. The EP (1997) still recommends the use of likely product contaminants, but there is no mention of extended sampling or repeated inoculation. The moves towards international harmonization mitigate against this kind of flexibility, but these, and the other test extensions or modifications which are considered below, can provide additional valuable information to the manufacturer.

### 10.3.1 *The use of additional test organisms and more precisely defined cultural conditions*

In the EP (1997) and in all editions of the USP up to and including 1995, there was a recommendation in the preservative efficacy tests that wild strains which might arise during manufacture or use of the product should be employed as test organisms in addition to the obligatory strains. The main argument in favour of this approach is that it provides the opportunity to use organisms which are realistic in terms of their preservative resistance. The pharmacopoeial strain of *Pseudomonas aeruginosa*, for example, is often found to be the most susceptible of the three bacteria used for testing purposes, and it is far more sensitive than the numerous reports of isolation of this species from preserved products would lead one to expect. Problems arise, however, when non-culture collection organisms are employed:

- Some manufacturers have used organisms isolated from previously spoiled batches of the product in question for this purpose on the basis that they represented realistic preservative-resistant strains. However, such cultures need to be maintained on preservative-supplemented media to avoid loss of resistance during repeated subculture. Cowen and Steiger (1976) showed that micro-organisms isolated from spoiled products often failed to establish themselves in a non-contaminated sample of the same preserved product, even though the organism in question was only subjected to a single overnight growth in a nutrient medium.

- 'In-house' strains are, by definition, not universally available, and so comparisons between different laboratories testing the same product may be difficult. Furthermore, if raw material or environmental isolates are used, these may change with time, so that successive batches may give different test results even within a single laboratory if the panel of test strains always reflects current contaminants.

Owing to these difficulties additional organisms are not recommended in the current USP test (1998) and the desire for international harmonization makes it unlikely that they will be recommended in future editions of the EP test (Dabbah, 1997).

### 10.3.2 *Mixed cultures, repeated challenges and variable inoculum concentrations*

It is well established that the growth of one organism might facilitate or impede the simultaneous or subsequent growth of another, either by the production of metabolic products acting as growth factors or toxins, or by organisms competing with each other for essential nutrients. The use of mixed cultures, therefore, would introduce the variable of population dynamics into the challenge test and make it a more realistic test since micro-organisms do not exist as pure cultures in nature (Brannan, 1995). This also applies to repeated inoculations, since some products are exposed to repeated challenge during use. However, more and more products are now packed as single-dose units, or the packaging restricts the opportunities for microbial ingress, e.g. creams packed in tubes rather than tubs or jars, and pump-action tubs replacing those with open tops. Nevertheless, it is still the case that topical products are more vulnerable to repeated microbial challenge than most other product categories, and in this respect pharmaceutical creams have much in common with many cosmetics.

A collaborative study was carried out by the Cosmetic, Toiletry and Fragrance Association (CTFA, 1981) Preservation Subcommittee to determine whether there was any advantage in using repeated product challenges rather than a single challenge in assessing preservative performance in cosmetic formulations, and it was concluded that repeated challenges were of value in detecting marginally preserved products. The CTFA (1993) recommends at least one re-challenge. Multiple inoculations may be criticised on the grounds that repeated challenge dilutes out the preservative within the formulation, and that increasing the organic bioburden can lead to a reduction in effective biocide by adsorption. This would be of more significance with biocides of high concentration exponent or when low initial concentrations are used.

The USP and EP require a challenge level of $10^5 - 10^6$ cells per ml within the product, and it is argued that such a level is necessary in order to determine accurately 3 log cycles of kill. A high concentration of cells also makes it more likely that the population contains mutant cells of unusually high preservative resistance. These may survive the effects of the preservative, become acclimatized and subsequently outgrow, so the use of an inoculum containing such cells might be regarded as testing for the 'worst case' scenario. However, in normal usage a product is unlikely to be challenged by such high concentrations of micro-organisms which may swamp the preservative as it adsorbs onto the cell surface giving, perhaps, artificially poor performance data. Despite this unrealistic element, all the current pharmacopoeial tests and those used in the cosmetics industry recommend high inocula, and no official consideration has been given in recent years to alternative inoculum levels.

### 10.3.3 *Rapid methods*

Conventional preservative efficacy testing methods are slow and labour-intensive; thus, alternative procedures have been recommended and recently reviewed (Hodges, 1998). These can generally be divided into two categories. First, methods which determine the initial rate of kill over a short period of exposure of the test organism to the preservative and use this value to predict the levels of survivors which would arise after longer periods. Alternatively, the standard 28-day sampling period is retained, but the effort associated with each sample is reduced by the use of 'new technology' such as impedance methods as a substitute for conventional viable counts. Both approaches are suitable when screening a relatively large number of preservative systems for a new product and the resource implications of assessing them all by conventional methods are unacceptable; preservative efficacy data generated by rapid methods are unlikely to be acceptable for product registration purposes, but may be found acceptable in developmental pharmaceutics.

The use of D-values (decimal reduction times) to extrapolate long-term performance from short-term exposure data is a common approach which has been recommended by a number of workers (Orth and Breuggen, 1982; Akers *et al.*, 1984). It relies, however, on an assumption of first-order kill kinetics (a plot of log of percentage survivors against time being linear), and has been criticized by Brannan (1995) since at the low biocide concentrations used in preservative systems the microbial kill might be expected to follow second-order kinetics. Sutton *et al.* (1991) found non-linear inactivation curves for contact lens disinfecting solutions, and likewise considered D-values to be an inappropriate method for determining biocidal performance in these solutions. Probably the best rapid methods using

**Table 10.7** Anticipated changes in European preservative efficacy tests

| Test feature | Pharmaceutical Discussion Group proposal* contrasted with EP (1997) |
|---|---|
| Environmental isolates | No environmental isolates. Previously: other 'likely contaminants' |
| Culture maintenance | Not more than five passages from culture collection. Previously: 'subcultures kept to a minimum' |
| Inoculum storage | Use within 8 h when stored at 2–8°C. Previously: use 'immediately' |
| Inoculum percentage | 0.5–1.0% by weight or volume. Previously: ≤1% of product volume. |
| Sampling times | Time zero sample eliminated |
| Performance criteria | B criterion for fungi reduced from 1 log kill to no increase at 14 days |
| | B criterion for bacteria reduced from 3 log to 2 log kill at 14 days |
| Definitions adopted | Introduction of 'no increase' = not more than 0.5 log greater than previous value measured. |

* From Dabbah (1997).

traditional microbiological techniques are the rapid screen test described by Farrington *et al.* (1994) and Fels' automated personal computer controlled test (Fels, 1995).

Connolly *et al.* (1993, 1994) have compared three rapid microbiological methods for evaluating preservative efficacy of pharmaceuticals and cosmetics. They found that impedance methods represented a potential alternative to conventional preservative efficacy tests, but that at their level of development at that time ATP determinations and direct epifluorescence methods could not be considered as satisfactory. Zhou and King (1995) have also reported favourable results with an impedimetric method for the rapid screening of cosmetic preservatives.

## 10.4 Future developments

The EP (1997) and USP (1998) testing protocols are now very similar, but it is still not possible to conduct a test using a procedure which simultaneously satisfies the two pharmacopoeias because of the different suspending media recommended for the inocula. The indications are, however, that the process of international harmonization will in due course resolve these differences if the recommendations shown in Table 10.7 are adopted.

Currently, there are no indications that the significant differences in performance criteria between the EP and USP are likely to be reconciled.

## References

AKERS, M.J., BOAND, A.V. and BINKLEY, D.A. (1984) Preformulation method for parenteral preservative efficacy evaluation. *Journal of Pharmaceutical Sciences*, **73**, 903–5.

AL-HITI, M.M. and GILBERT, P. (1980) Changes in preservative sensitivity for the USP antimicrobial agents effectiveness test micro-organisms. *Journal of Applied Bacteriology*, **49**, 119–26.

AL-HITI, M.M. and GILBERT, P. (1983) A note on inoculum reproducibility: solid versus liquid culture. *Journal of Applied Bacteriology*, **55**, 173–6.

BAIRD, R.M. (1995) Preservative efficacy testing in the pharmaceutical industries. In: BROWN, M.R.W. and GILBERT, P. (eds), *Microbiological Quality Assurance: A Guide Towards Relevance and Reproducibility of Inocula*. CRC Press, New York, pp. 149–62.

BRANNAN, D.K. (1995) Cosmetic preservation. *Journal of the Society of Cosmetic Chemists*, **46**, 199–220.

*British Pharmacopoeia* (1999) The Stationery Office, London.

CONNOLLY, P., BLOOMFIELD, S.F. and DENYER, S.P. (1993) A study of the use of rapid methods for preservative efficacy testing of pharmaceuticals and cosmetics. *Journal of Applied Bacteriology*, **75**, 456–62.

CONNOLLY, P., BLOOMFIELD, S.F. and DENYER, S.P. (1994) The use of impedance for preservative efficacy testing of pharmaceuticals and cosmetic products. *Journal of Applied Bacteriology*, **76**, 68–74.

COWEN, R.A. and STEIGER, B. (1976) Antimicrobial activity – a critical review of test methods of preservative efficacy. *Journal of the Society of Cosmetic Chemists*, **27**, 467–81.

CTFA (1993) The determination of preservative adequacy of water-miscible cosmetic and toiletry formulations. *Cosmetic, Toiletry and Fragrance Association Microbiology Guidelines*, CTFA, Washington.

CTFA Microbiology Committee (1981) A study of the use of rechallenge in preservation testing of cosmetics. *CTFA Cosmetic Journal*, **13**, 19–22.

DABBAH, R. (1997) Harmonization of microbiological methods – a status report. *Pharmacopoeial Forum*, **23**, 5334–44.

DABBAH, R., CHANG, W.W. and COOPER, M.S. (1996) The use of preservatives in compendial articles. *Pharmacopoeial Forum*, **22**, 2696–704.

DAVISON, A.L. (1996) Preservative efficacy testing of pharmaceuticals, cosmetics and toiletries and its limitations. In: BAIRD, R.M. and BLOOMFIELD, S.F. (eds), *Microbial Quality Assurance in Cosmetics, Toiletries and Non-Sterile Pharmaceuticals*, 2nd edn. Taylor & Francis, London, pp. 187–98.

DENYER, S.P. and WALLHAEUSSER, K.-H. (2001) Antimicrobial preservatives and their properties. In: DENYER, S.P. and BAIRD, R. (eds), *Guide to Microbiological Control in Pharmaceuticals*. 2nd edn. Ellis Horwood, London, In press.

*European Pharmacopoeia* (1997) 3rd edn. Council of Europe, Strasbourg, pp. 286–7.

FARRINGTON, J.K., MARTZ, E.L., WELLS, S.J., ENNIS, C.C., HOLDER, J., LEVCHUK, J.W., AVIS, K.E., HOFFMAN, P.S., HITCHENS, A.D. and MADDEN, J.M. (1994) Ability of laboratory methods to predict in-use efficacy of antimicrobial preservatives in an experimental cosmetic. *Applied and Environmental Microbiology*, **60**, 4553–8.

FELS, P. (1995) An automated personal computer-enhanced assay for antimicrobial preservative efficacy testing by the most probable number technique using microtitre plates. *Pharmaceutical Industry*, **57**, 585–90.

GILBERT, P. and BROWN, M.R.W. (1995a) Continuous culture of micro-organisms in liquid media: effects of growth rate and nutrient depletion. In: BROWN, M.R.W. and GILBERT, P. (eds), *Microbiological Quality Assurance; A Guide Towards Relevance and Reproducibility of Inocula*. CRC Press, New York, pp. 31–48.

GILBERT, P. and BROWN, M.R.W. (1995b) Some perspectives on preservation and disinfection in the present day. *International Biodeterioration and Biodegradation*, **36**, 219–26.

GILBERT, P., CAPLAN, F. and BROWN, M.R.W. (1991) Centrifugation injury to micro-organisms. *Journal of Antimicrobial Chemotherapy*, **27**, 550–1.

HODGES, N.A. (1998) Assessment of preservative activity during stability studies. In: MAZZO, D. (ed.), *International Stability Guidelines*. Interpharm Press, Buffalo Grove, IL, pp. 115–45.

HODGES, N.A. and DENYER, S.P. (1996) Preservative testing. In: SWARBRICK, J. and BOYLAN, J.C. (eds), *Encyclopedia of Pharmaceutical Technology. Volume 13.* Marcel Dekker, New York, pp. 21–37.

HODGES, N.A., DENYER, S.P., HANLON, G.W. and REYNOLDS, J.P. (1996) Preservative efficacy tests in formulated nasal products: reproducibility and factors affecting preservative activity. *Journal of Pharmacy and Pharmacology*, **48**, 1237–42.

HUGO, W.B. and RUSSELL, A.D. (1998) Evaluation of non-antibiotic antimicrobial agents. In: HUGO, W.B. and RUSSELL, A.D. (eds), *Pharmaceutical Microbiology*, 6th edn. Blackwell Science, Oxford, pp. 229–55.

*Japanese Pharmacopoeia* (1996) 13th edn. The Society of Japanese Pharmacopoeia, Tokyo, pp. 1069–71.

LIVINGSTONE, D.L., HANLON, G.W. and DYKE, S. (1998) Evaluation of an extended period of use for preserved eye drops in hospital practice. *British Journal of Ophthalmology*, **82**, 473–5.

ORTH, D.S. and BRUEGGEN, L.R. (1982) Preservative efficacy testing of cosmetic products. Rechallenge testing and reliability of the linear regression method. *Cosmetics and Toiletries*, **97**, 61–5.

*Pharmacopoeial Forum* (1992) Pharmacopoeial Previews. ⟨51⟩ Antimicrobial effectiveness testing. Vol. 18, 3048–52.

Pharmeuropa (1994) International harmonization. VIII.14 Efficacy of antimicrobial preservation. Vol. 6, 387–8.

RUSSELL, A.D. (1981) Neutralization procedures in the evaluation of bactericidal activity. In: COLLINS, C.H., ALLWOOD, M.C., BLOOMFIELD, S.F. and FOX, A. (eds), *Disinfectants: Their Use and Evaluation of Effectiveness*. Academic Press, London, pp. 45–59.

SUTTON, S.V., FRANCO, R.J., PORTER, D.A., MOWREY-MCKEE, M.F., BUSSCHAERT, S.C., HAMBERGER, J.F. and PROUD, D.W. (1991) D-value determinations are an inappropriate measure of disinfecting activity of common contact lens disinfecting solutions. *Applied and Environmental Microbiology*, **57**, 2021–6.

*United States Pharmacopoeia* (1995) 23rd rev. ⟨51⟩, Antimicrobial Preservatives – Effectiveness. US Pharmacopoeial Convention, Rockville, MD, p. 1681.

*United States Pharmacopoeia* (1998) Eighth supplement to 23rd edn. Pharmacopoeial Convention, Rockville, MD, pp. 4293–5.

*United States Pharmacopoeia* (1999) Tenth supplement to 23rd edn. Pharmacopoeial Convention, Rockville, MD, pp. 5063–5.

ZHOU, X. and KING, V.M. (1995) An impedimetric method for rapid screening of cosmetic preservatives. *Journal of Industrial Microbiology*, **15**, 103–7.

# 11

# Microbiological Assay of Antibiotics in Pharmaceutical Preparations

COLIN THOMPSON

*GlaxoWellcome, Temple Hill, Dartford, Kent, DA1 5AH, United Kingdom*

## 11.1 Introduction

The antibiotic bioassay, still required by major pharmacopoeias such as the British (BP 1999), European (EP 1997) and United States (USP 1995), is the major survivor of a bygone era in which a biological assay of a natural substance was the ultimate measure of its suitability for use. Today, many of the original pharmacopoeial biological tests are no longer required. The assays for digitalis, once requiring the use of pigeons (USP) or guinea pigs (BP), and insulin, requiring the use of rabbits (USP) or mice (BP), are now replaced by more reliable chemical methods such as high-performance liquid chromatography (HPLC). These replacements have been facilitated mainly by an increase in purity of the active substances, although the desire to avoid animal use has also played a major part. The assay of antimicrobial agents in body fluids has also, traditionally, been performed biologically but here again, alternative methods such as HPLC have proved more acceptable and have been widely adopted. A major advantage in the clinical situation is that chemical determinations are rapid, although the equipment required is often expensive and can incur substantial maintenance costs.

The biological determination of antibiotic potency in pharmaceutical preparations is, however, unchanged in principle and little changed in practice when compared with procedures used in the 1950s. Antibiotics used in medicine are often not pure substances, but are produced mainly from well-defined large-scale processes and may comprise a variety of closely related components that individually may display considerable differences in biological activity and availability. For example, the multiple components of polymyxin B which are revealed by HPLC are illustrated in Figure 11.1.

The antibiotic bioassay provides a single value (termed 'potency') for the overall biological activity of an antibiotic preparation. The activity or potency is quoted in terms of units of a recognized international standard, specifically defined and required by pharmacopoeias. For many antibiotics there is good correlation between their weight and international units; despite this, potencies are typically quoted in international units rather than weight. The World Health Organization, an agency

190

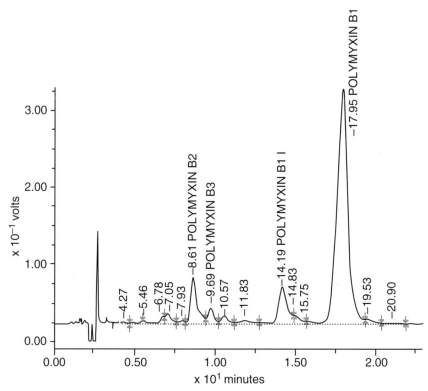

**Figure 11.1**　Segregation of components of polymyxin B sulphate using high-performance liquid chromatography (HPLC).

of the United Nations, undertakes the provision of international antibiotic standards, establishing definitive units of activity by means of extensive collaborative study.

Segregation and quantification of antibiotic components by chemical methods such as HPLC, although precise, may not provide a true indication of biological activity, and attempts to correlate antibiotic bioassay results with those from chemical methods have usually proved disappointing. Bioassays continue to play an essential role in the manufacture and quality control of antibiotic medicines, and still demand considerable skill and expertise to assure success. This chapter describes the two main approaches to antibiotic assay, agar diffusion and broth turbidity (turbidimetric assays), and explains the many factors which may influence assay performance and the reliability of the result.

## 11.2　Principles of antibiotic assay

Both agar diffusion (plate assays) and turbidimetric assays permit an estimate of antibiotic potency through direct comparison of a test antibiotic with an approved, well-calibrated, reference substance. Agar diffusion is usually the method of choice whenever the nature of an antibiotic permits (in particular, an antibiotic must be water-soluble for assay by diffusion). In the plate method, active components of antibiotics diffuse, during a defined period at optimum temperature, through an appropriate medium which is seeded with a sensitive organism. The medium is gelled by agar, a carbohydrate extract of certain seaweeds, whose chain-like molecules provide a framework in which water forms a continuous capillary network through which the antibiotic components diffuse. Where the antibiotic concentration is sufficiently high, microbial growth is inhibited, but where the concentration is insufficient, microbial growth is displayed. The border of these effects defines a clear boundary and a zone of inhibition is created (Figure 11.2; see also Figure 11.4). With most antibiotics, a linear relationship exists between the diameter or area of this zone of inhibition and the logarithm of the antibiotic concentration producing the zone. Generally, the more potent the antibiotic, the larger is the zone of inhibition.

The diffusion method has a number of variations which all deploy the same principle. Assays may be conducted in which the antibiotic solutions are placed in wells cut into the media, in cylinders balanced on the surface, or absorbed into discs placed onto the agar. The basic approach may be varied to determine: antibiotic potencies in finished pharmaceutical preparations, their intermediates or raw materials; the sensitivity of pathogens in clinical material to various antibiotics; or the concentration profile of an administered antibiotic in body fluid. Agar diffusion is also the definitive method prescribed by regulators when detecting the presence of sensitizing antibiotics such as penicillin in suspect pharmaceutical formulations, as this has proved to be more sensitive than HPLC.

Turbidimetric assessment is used less widely than agar diffusion, although it might be considered to resemble more closely the clinical situation where growth of an organism in a liquid culture is directly challenged by antibiotic treatment. Its major use is for those antibiotics such as gramicidin, which are poorly soluble in water and have to be prepared in another solvent such as ethanol. Here, a nutrient broth is inoculated with an organism sensitive to the test antibiotic and dispensed into tubes containing standard and test antibiotic solutions. Incubation permits the organism to grow

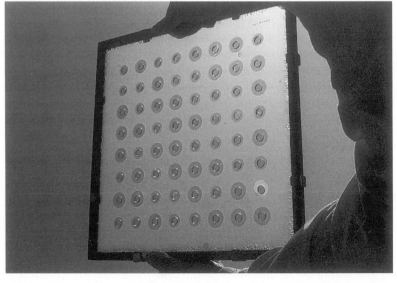

**Figure 11.2**  Antibiotic agar diffusion assay showing zones of inhibition obtained on a large glass/aluminium plate.

**Table 11.1**   Relative merits of different antibiotic assay methods

|  | Agar diffusion | Turbidimetric | HPLC |
|---|---|---|---|
| Equipment | Basic | Basic | Specialized |
| Training | Moderate | Moderate | Moderate |
| Technical skills | High | High | Moderate |
| Regulatory status | Acceptable | Acceptable | Must show equivalence |
| Equivalence to pharmacopoeial method | Full | Full | Difficult to prove |
| Result availability (h) | 24 | 5 | $\leq 1$ |
| Cost of equipment | Low | Low | High |
| Method precision | Moderate | Variable or poor | Good |
| Relationship to clinical use | Moderate | High | Poor |
| Reliability/ruggedness | Moderate | Variable | Good |
| Susceptibility to change in test conditions | Moderate | High | Low |

quickly and develop a turbidity which is inversely proportional to the logarithm of antibiotic concentration inhibiting the growth.

In theory, there are several possible approaches which could be used to determine the activity of an antibiotic preparation, and it is perhaps for historic rather than sound scientific reasons that biological methods continue to be favoured by pharmacopoeias and regulators. HPLC would certainly offer advantage in terms of speed and precision, but concerns remain regarding the significance of results obtained by this method in relation to the clinical use of mixed-component antibiotics. Agar diffusion and turbidimetric determinations are well established and both permit potency determinations of pharmaceutical products which are acceptable to regulators; agar diffusion has proved the most rugged in use. The merits of the different assay methods are summarized in Table 11.1.

## 11.3   Practical aspects of antibiotic assays

Antibiotic assays are not as technically simple as they may appear, and readily suffer variation from subtle changes in environment (temperature, humidity, thickness of media, pH) or technical approach. Considerable practice and attention to detail is required if consistent, reproducible results are to be achieved.

### 11.3.1   *Preparation of diffusion plates*

Diffusion assays are performed in clean, sterile plates varying in size and materials from Petri dishes to large glass or glass/aluminium rectangular plates (see Figure 11.2). Where occasional assays are performed, disposable plates are ideal (section 11.5); their increased price is offset by reduced cleaning costs (such plates must still be rendered safe before disposal). However, when frequent assays are required, the cost of disposable plates can be prohibitive and recycling of glass

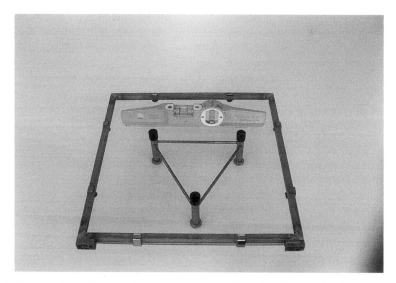

**Figure 11.3**   Levelling of assay plates using a screw-foot tripod and spirit level.

plates becomes cost-effective. Where plates are used repeatedly it is essential that the cleaning process is sufficiently rigorous to remove all traces of antibiotic used in previous assays, as well as any disinfectant or detergent residues. An inefficient cleaning regime will quickly lead to carry-over problems and inaccurate results in subsequent assays.

Plastic Petri dishes are usually sterilized by irradiation, while glass plates are either autoclaved, oven-sterilized or flamed prior to use. Autoclaving of glass/aluminium plates can lead to a failure in the seal between the glass and aluminium. This can be remedied through a thread of molten agar which is applied around the interface by sterile Pasteur pipette, and allowed to set before the plate is filled with medium.

Diffusion characteristics are affected by the thickness of agar and this must, therefore, be uniform throughout the plate. In order to achieve an even distribution the plates must be levelled before medium is introduced; a simple screw-foot tripod and spirit level may be employed for this purpose (Figure 11.3).

### 11.3.2   *Test organisms*

The basic requirements for a test organism are that it should be non-pathogenic, sensitive to the action of the antibiotic under assessment, and capable of rapid growth. If a mixed antibiotic formulation is involved, the micro-organism of choice should be sensitive only to that antibiotic whose potency is being determined. Should this not be possible, use of an appropriate substance to neutralize the unwanted antibiotic is an alternative strategy; a list of such agents is given by Andrews (1999). Appropriate micro-organisms for the assay of most antibiotics are well defined in the BP (1998), EP (1997) and USP (1995). Their purity, viability and 'vigour' all contribute to the success of an assay, and careful preparation of a pure test culture is essential. Master cultures obtained from established culture collections such as the National Collection of Type Cultures (NCTC) or National Collection of Industrial and Marine Bacteria (NCIMB) in the UK, or the American Type Culture Collection

(ATCC) in the USA, may be used to prepare working cultures suitable for use in routine assays. It is advisable to perform purity and identity checks on both master and working cultures (see Chapter 5) and, in order to avoid the possibility of significant mutation from the original, there should be no more than five passages (subcultures) between the master culture obtained from a national collection and the laboratory working culture used to seed the assay medium.

For routine use, non-sporing cultures may be conveniently maintained on agar slopes stored between 2°C and 8°C. Alternatively, aliquots of suspended culture in growth medium supplemented with 10% (w/w) glycerol may be dispensed into vials and stored under liquid nitrogen at about −196°C for up to 6 months. For maximum recovery from deep freezing, vials should be thawed quickly at 43°C and not simply be allowed to achieve room temperature by equilibration. Re-freezing of thawed vials should be avoided, as loss of organism viability or 'vigour' is likely to result. Pharmacopoeial procedures recommend spore inocula for assays of several antibiotics. Spore suspensions can be produced by harvesting growth from a *Bacillus* culture, e.g. *Bacillus pumilus*, *Bacillus subtilis* or *Bacillus cereus* on Antibiotic Assay Medium No. 1 (as described in the pharmacopoeias) incubated for 7 days at 35–37°C (*Bacillus pumilus* or *Bacillus subtilis*) using small, sterile glass beads and sterile distilled water. However, for *Bacillus cereus* a better harvest might be achieved at a slightly lower temperature, e.g. 30°C. The suspension should consist mainly of spores, but any residual vegetative cells may be eliminated by heating at 70°C for 30 min. An aqueous spore suspension prepared in this way is extremely rugged and could remain viable when stored at 2–8°C for several years.

Assay cultures should be standardized to provide reproducible, clear, well-defined zones of inhibition of satisfactory diameter under the specific conditions of the assay. The extent to which zone diameter may be influenced by the concentration of the inoculum is illustrated in Figure 11.4. Once a satisfactory concentration has been determined, its turbidity may be used as a means of standardizing future lots.

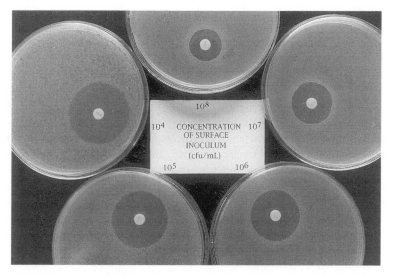

**Figure 11.4** The influence of inoculum concentration on benzylpenicillin inhibition zone diameter. Petri dishes surface-inoculated with 0.2 ml *Bacillus subtilis* spore suspensions at concentrations (clockwise from the top) $10^8$ to $10^4\,\text{ml}^{-1}$.

### 11.3.3  *Media and diluents*

The choice of appropriate media and diluents is of enormous importance in antibiotic assays. Suitable media are required both for the preparation of test organism
inocula and for the conduct of the assay. Media must provide adequate nutrition as
well as physical conditions (e.g. pH) suitable to promote growth of the test organism
while also facilitating diffusion of the antibiotic. There is a vast choice of potential
media formulae, but guidance regarding the most suitable is again provided by
pharmacopoeias.

Most culture maintenance and assay media are available commercially in dehydrated form. Simple reconstitution of these in high-quality demineralized or distilled
water provides the most reliable assay media. Care exercised during medium
preparation will contribute significantly to the reliability of assays. Media must be
sterilized but not overheated, as both nutritive and gelling properties may be adversely
affected. If heat-sensitive additives (e.g. neutralizers) are employed they may be better
sterilized by filtration (0.2 µm) and added aseptically prior to use. Before use, each
batch must be checked for sterility and also for growth promotion characteristics by
challenging aliquots with small numbers (10–100 c.f.u.) of selected organisms. The pH
is also important both for growth promotion of the test organism and diffusion
characteristics of the test antibiotic. Media should be stored under defined conditions
(refrigerated or room temperature and in the absence of light) for validated periods.
When required for use they should be liquefied, then cooled and equilibrated in a
water bath to a temperature of 46–50°C, at which point the prepared culture may be
added and mixed thoroughly. Seeded medium is poured into the assay plate, taking
care to avoid the presence of air bubbles, to a depth of between 2 and 5 mm.
Alternatively, the medium may consist of two layers, separately poured, with only
the upper layer being inoculated; this procedure may afford greater sensitivity than
a single layer plate and so be of value in detecting low concentrations of cross-
contaminating antibiotics, e.g. penicillin. The depth of agar must be uniform
throughout the plate.

### 11.3.4  *Reference standards*

Reference standards used in the microbiological assay of antibiotics are substances
whose activity has been precisely determined, perhaps by means of international
collaborative studies. In such studies the issuing authority invites a selection of
competent laboratories from different countries to determine the potency of a proposed new reference standard, usually in terms of that which is currently used. The
assigned potency for the new reference standard represents a detailed statistical
assessment and combination of all the results achieved. Accuracy of the microbiological assay is dependent upon the correct use of these standards and care should be
taken with their storage and preparation.

Some laboratories choose to use two independent standard preparations from
separate weighings in each assay, and will reject the assay if the results from the two are
not in good (e.g. 10%) agreement.

Reference antibiotic standards are readily available (see section 11.5) from the
following sources:

- WHO (International Standard)
- PhEur (Chemical Reference Substance)
- USP (Reference Standard)

All are supplied in sealed containers protected from moisture and air. They should be stored under these same conditions at specified temperatures and, under no circumstances, above 8°C. WHO and PhEur standards are supplied with certificates and batch numbers, but with no indication of expiry. It is particularly important, therefore, that their receipt is documented and that their use and storage is strictly controlled. Ideally, use should be restricted to a maximum of one year from receipt. USP standards are supplied as 'working standards' and are subject to frequent replacement by the authority. Information listing 'current' batch numbers for antibiotic reference standards is provided through the USP 'Pharmacopoeial Forum'. In contrast, International Standards (WHO) are regarded as 'primary standards' against which 'working standards' are calibrated, and their batch numbers rarely change (for instance a reference standard for bacitracin remained unchanged for 30 years). All dilutions of reference standards should be prepared fresh on the day of use. Any weighing should be performed using a fully validated and calibrated analytical balance, accurate to four decimal places. Weighings should be performed quickly in a low-humidity environment, as many antibiotics (e.g. polymyxin B sulphate) are capable of moisture absorption, and a reference substance affected by moisture absorption during weighing will give rise to false high potencies for the test material in an assay. United States antibiotic working standards must be dried in a vacuum oven prior to weighing. Dilutions should be prepared using clean and accurately calibrated glassware (of at least Grade B quality).

### 11.3.5 Test preparations

Antibiotic-containing materials being examined by microbiological assay are conveniently referred to as 'unknown' or 'test'. Their preparation is varied; both antibiotic test and reference materials must be completely dissolved in a suitable solvent and then diluted to required concentrations. Phosphate buffers providing pH values, after sterilization, between 4.5 (e.g. for tetracyclines) and 8.0 (e.g. for aminoglycosides) are most frequently employed as diluents, although purified water may be more appropriate for some antibiotics. Others may require complex extraction from formulated products (including the removal of substances which could interfere with the microbiological assay). Difficult extractions such as those from creams may be performed in a 'Stomacher' (see section 11.5). Here, the cream in an extractant buffer is continuously agitated within a flexible plastic container. Extraction procedures should be fully validated to show satisfactory recovery of the antibiotic. All weighings of the unknown or test must be performed using a fully validated and calibrated analytical balance. Volumetric measurements for extraction and dilution should be achieved using clean and accurately calibrated glassware (of at least Grade B quality).

### 11.3.6  Conduct of diffusion assays

Solutions of standards and unknowns are applied to assay plates by one of the following methods:

- in cavities (wells) of about 8 mm diameter punched from the agar. A specially fabricated plate cutter (Figure 11.5) may facilitate this process. The best defined and cleanest cavities will be obtained if the agar is cooled for a short period at 2–8°C prior to cutting.

- on absorbent discs or in porcelain or stainless steel cylinders of about 6 mm diameter, balanced on the agar surface. Even spacing and balancing of these cylinders is a delicate operation and is best achieved using a mechanical guide. The cylinders must be clean and free from all residues of previous use. Occasional washing in nitric acid (2 M) or chromic acid (200 g sodium dichromate, 100 ml water, 1500 ml sulphuric acid) is ideal, but must be performed with safety in mind.

In Europe, wells are generally preferred as they are the easiest to prepare and maintain, and are the most suitable for use in large scale assays. In contrast, cylinders are favoured in the USA, mainly because their use is prescribed by the USP; absorbent discs are now rarely employed.

In large plate assays the wells or cylinders should be arranged in a rectangular grid pattern in which none should be closer than 25 mm from the edge, nor from adjacent wells or cylinders. Advantage should be taken of statistical design (e.g. Latin Square or randomized block dosing arrangement) in order to reduce variations such as those that might arise from small fluctuations in agar depth, culture concentration or incubation temperature within the plate. The volume of solution added to each well, cylinder or disk must be constant (typically 100 µl) and should be applied from a

**Figure 11.5**  Semi-automatic plate cutter. Smooth pressure on the handle lowers 64 stainless steel cutters onto the agar-filled diffusion plate.

calibrated pipette or syringe (ideally to an accuracy of ±2%). Usually, three different dilutions of standard and unknown in a constant dose ratio must be used on each plate; the doses (concentrations) selected must be shown to produce a linear calibration plot of log antibiotic concentration against zone diameter or area. Pharmacopoeias do, however, allow for this to be reduced to two dilutions provided that extensive experience and validation using multiple calibrator solutions supports a satisfactory linear regression between the chosen doses. Each test preparation is given an 'assigned potency' based on the 'expected result', and solutions of the unknown are prepared to be identical in strength to the standard solutions, on the basis of the 'assigned potency'. When performed in Petri dishes, standards and unknowns are placed in alternate wells or cylinders in a manner which avoids the potential for interaction of the more concentrated solutions. Larger plates make use of an appropriate statistical design. Assay design, and hence number of replicate doses, will be dependent on the degree of precision required and the number of test samples to be assayed within the same plate. Examples of a high-precision assay (ideal when the assay is required only on an occasional basis for discrete samples) and a 'medium-precision comparative assay' (preferred when several batches or part batches of a product need to be compared in the same assay) are given in Tables 11.2 and 11.3.

Once the addition of solutions is complete, the assay plates may be 'incubated' at room temperature or even at 4°C for at least 1 h. This period will minimize the effect of time variation between additions while also permitting diffusion of the antibiotic components to commence when the test organism is relatively dormant, thus increasing the size and possibly the clarity of inhibition zones and improving regression of the dose response.

Plates should be incubated at a temperature between 32°C and 39°C (depending upon the assay organism) for an optimum period (usually 12–18 h) as determined during the development and validation of the assay. Different incubation regimes may be required for different antibiotics and associated organisms involved. It is important to determine a period of incubation which provides the best balance between antibiotic diffusion and culture growth. The intention is to achieve well-defined zones of inhibition of adequate size and showing good regression between the dose levels chosen. After incubation, a typical plate will appear opaque due to the uniform growth of micro-organisms within the agar. Clean annular zones will be discernible around

**Table 11.2** A high-precision assay design (8 × 8 Latin Square)

Dosing arrangement of wells

| | | | | | | | |
|---|---|---|---|---|---|---|---|
| 3 | 7 | 6 | 8 | 2 | 5 | 4 | 1 |
| 8 | 5 | 3 | 6 | 4 | 1 | 7 | 2 |
| 4 | 6 | 2 | 7 | 1 | 8 | 5 | 3 |
| 1 | 3 | 4 | 5 | 8 | 7 | 2 | 6 |
| 2 | 1 | 8 | 4 | 5 | 6 | 3 | 7 |
| 7 | 8 | 5 | 3 | 6 | 2 | 1 | 4 |
| 6 | 4 | 1 | 2 | 7 | 3 | 8 | 5 |
| 5 | 2 | 7 | 1 | 3 | 4 | 6 | 8 |

Where:  1 and 3 Test high
2 and 4 Test low
5 and 7 Standard high
6 and 8 Standard low

**Table 11.3**  A medium-precision assay design using 'check standards'

Dosing arrangement of wells

| 4  | 13 | 10 | 5  | 1  | 7  | 16 | 12 |
|----|----|----|----|----|----|----|----|
| 8  | 11 | 3  | 14 | 9  | 6  | 2  | 15 |
| 15 | 2  | 6  | 9  | 8  | 3  | 11 | 14 |
| 12 | 16 | 7  | 1  | 5  | 13 | 10 | 4  |
| 14 | 3  | 11 | 8  | 15 | 2  | 6  | 9  |
| 5  | 10 | 13 | 4  | 12 | 16 | 7  | 1  |
| 1  | 7  | 16 | 12 | 4  | 10 | 13 | 5  |
| 9  | 6  | 2  | 15 | 14 | 11 | 3  | 8  |

Where:  1 Standard A high
2 Standard A low
3 Standard B high
4 Standard B low

5, 7, 9, 11, 13, 15      Tests (1 to 6) high
6, 8, 10, 12, 14, 16     Tests (1 to 6) low

Standard A is used as the definitive reference.
Standard B is used as an independent check of Standard A

the wells or cylinders where the antibiotic has diffused through the agar and inhibited growth; zone diameters of 14–28 mm are considered to be most conducive to the attainment of reliable results (Hewitt and Vincent, 1989).

Precise measurement of the zone of inhibition is essential, and may be performed manually or automatically using optical means or with callipers. Most commonly used are projection systems in which zones are projected onto a screen such that their enhanced diameters are more easily measured. Results might be manually achieved and recorded or, in some cases, captured electronically. Comprehensive projection systems which includes a plate reading device, recorder and assay analysis software may be obtained (see section 11.5). An alternative, which is growing in popularity, is the measurement of zone area by image analysis where electronic comparison of 'light' and 'dark' provide a measure of the area of inhibition. Image analysis is best suited to assays providing good discrimination between the confluent growth of an organism (e.g. *Micrococcus luteus*) and the medium base. It is less suitable for *Bacillus* species, which often present as a granular growth. The measurement of diameter should be taken through the centre of the zone which, provided that a good technique was employed earlier in the assay, should present as a perfect circle. Distorted zones may result from poorly cut wells or poor addition of solutions and should be excluded from the assay (i.e. treat as 'missing value' in the calculation of potency). When manual measurement is used it should be performed 'blind' in order to avoid subconscious bias (i.e. without reference to the treatment administered). All measuring devices should be accurate and reproducible to at least 0.1 mm.

### 11.3.7  *Conduct of turbidimetric assays*

For turbidimetric assays, nutrient broth is inoculated with an appropriate test organism and dispensed immediately after preparation into a series of identical tubes each containing the same volume of either standard or unknown solutions. It is important that all tubes are clean, and if turbidities are to be measured direct from tubes, they should also be free from both blemishes or scratches. Typically, three levels of both standard and unknown test concentration are employed with about five replicates for each treatment. However, in order to establish the most suitable doses for routine use it might first be necessary to perform assays with a larger number of dose levels selected to achieve culture absorbance (optical density) values normally in the range 0.1 to 0.5. Two control tubes containing the same volume of antibiotic-free test diluent should be included to monitor the uninhibited rate of growth for the test organism. Both are filled with inoculated medium. One, the growth control, is untreated while the other, the negative control is further treated immediately with formaldehyde (0.5 ml of 35% in 10 ml broth) to prevent growth. These are used to set the optical apparatus used to measure growth; typically, wavelengths between 370–380 nm are employed (see also Chapter 4). The inoculated tubes are incubated, usually at 36–37.5°C, for about 3–4 h using either a thermostatically controlled water bath or dry air incubator. The apparatus chosen should be capable of bringing all tubes uniformly and quickly to the appropriate incubation temperature. At the end of incubation further growth is prevented by addition of formaldehyde or an alternative inhibitor, and opacities are read photometrically. Most reliable readings will be achieved using a flow-through cell, although satisfactory readings may also be made manually by transfer to cuvettes. Less reliable results are achieved by direct reading of the tubes in a nephelometer. Turbidity produced is normally inversely proportional to the logarithm of the antibiotic concentration [although alternative relationships may also give a straight line calibration, e.g. concentration on a linear scale against log of absorbance (Hewitt and Vincent, 1989)]

Turbidimetric assays are the most likely to be considered as candidates for automation. Systems are readily available for the preparation and dispensing of solutions and media (see section 11.5) and also for the automatic reading of turbidity. Results obtained can be automatically captured by a computer that is also programmed to perform the required potency and associated calculations. When considering automation, however, it should be noted that the turbidimetric assay is particularly sensitive to change. Great care must be taken in order to achieve consistency between assays. For example, surprising variations in growth characteristics of the test organism, *Enterococcus faecalis*, have been observed while assaying the antibiotic gramicidin in tubes of different construction (i.e. glass, polypropylene and expanded polystyrene). Growth for this organism was much slower in the glass tubes (assay length 4–4.5 h) than in the plastic variations (assay length 3–3.5 h), although assays conducted in glass still provided the best overall results with consistently low variation between replicates and accurate potency determinations.

### 11.3.8 *Calculations*

The results of an antibiotic potency assay consist of two numeric elements. One is a characteristic of the unknown sample, termed the 'estimate of potency', while the other is a measure of assay precision, termed the 'confidence or fiducial limit'. Both are obtained from appropriate mathematical treatment of the biologically generated data. Calculations are best performed using a validated software package which will challenge the validity of each assay, combine results from individual assays, and provide both potency and confidence limit determinations; such packages are available commercially. Appropriate mathematical models are described in detail in section 11.5.3 of the EP (1997) and ⟨111⟩ of the USP (1995). Fully worked numerical examples for a variety of different assay designs are to be found in: section 5.3.3.2 of the EP (1997), Hewitt and Vincent (1989) and Wardlaw (1985). Antibiotic assays must be designed such that their validity can be challenged by the model chosen to determine potency values. In Europe, a two- or three-dose parallel line model is usually favoured, while in the USA potencies are normally interpolated from a '1-level' assay where several replicates of a single dilution of an unknown are compared with a response curve obtained from five concentrations of a reference standard. Unfortunately, it is not easy to prove equivalence of these substantially different approaches. Consequently, companies intending to supply antibiotic products internationally are best advised to choose the approach most relevant to their markets. Products supplied to European markets should satisfy the EP method, while those supplied to US and Canadian markets should be subject to the USP approach.

The two- or three-dose parallel line design favoured in Europe requires that the log dose responses of unknown and standard solutions must be both parallel and linear over the range of the doses to be considered by the calculation. Often these conditions are satisfied only for a limited dose range, and the calculations employed must include appropriate checks to ensure that both rectilinearity and parallelism were satisfactorily achieved. Although it is possible to perform parallel line assays using just two dose levels, satisfactory rectilinearity must be demonstrated in an adequate number of experiments using the three-dose assay before the reduction can be justified.

Tests for validity based on a statistical probability of '$P = 0.05$' are used to verify that the required assay conditions have been met. Assay results are considered to be statistically valid only when:

- significant regression ($P < 0.01$) of the dose response is achieved (i.e. the response slope is well defined);

- linearity of the dose response is confirmed for the chosen range (probability of non-linearity $< 0.05$); and

- the unknown and standard solution dose responses are parallel (probability of non-parallelism $< 0.05$).

Where, for example, a lack of parallelism is identified when comparing an unknown and standard response, the two cannot be accepted (statistically) as representing the same active principle, and the assay is therefore invalid.

The potency of an 'unknown' calculated from any biological assay is more correctly described as an estimate of the 'true potency'. This is determined together with confidence (or fiducial) limits which are a measure of error variance for the assay. Confidence limits of $P = 0.95$ are quoted based on a 95% probability that they include the true potency. Simply, the confidence or fiducial limits are so computed that their upper and lower values would expect to enclose the true potency in 19 out of 20 valid assays.

In the United States '1-line' approach the concentration of an unknown is prepared with the intention that it will provide a response about mid-way when compared with a standard response curve. A log relative potency for each replicate of the unknown solution is determined against a curve computed from the various replicates of the standard dose response. The average of these determinations provides the final assayed log relative potency for the unknown.

Microbial determinations of antibiotic potency are subject to variation between separate assays. Consequently, at least two independent assays are required in order to provide a reliable estimate of potency for a given assay preparation. The calculated confidence limits for such assays would normally overlap, but in some cases the estimated potencies might differ significantly as indicated by the calculated standard error. In this case further assays will be required. A combined result achieved from a series of independent assays performed over a number of days would normally expect to provide a far more reliable estimate of potency than that achieved from a single, but larger assay.

The nature of the antibiotic assay does, at times, provide an abnormal response which might justify omission from calculation. Such rejection should, however, be used sparingly and with care in order to avoid a potential for bias in the assay. Rejection of abnormal results should be justified against appropriate statistical criteria such as the 'Test for Outliers' detailed in USP (1995). Occasional missing values, arising from such omissions or from 'accidents' occurring within the assay may be replaced statistically, thus retaining the balance of assay design using guidance provided in both the European and United States Pharmacopoeias.

## 11.4  Summary

Biological assays conducted by agar diffusion or turbidimetric methods are described in the major pharmacopoeias and, despite the increased popularity of HPLC and other chemical methods, bioassays are still preferred in a number of situations – particularly for antibiotics which consist of a number of active components possessing intrinsically different antimicrobial activities. Although simple in principle, microbiological assays of antibiotics represent a significant challenge in terms of the achievement of acceptable accuracy and precision, and reliable results are most likely to be forthcoming when careful attention is paid to the control of the various factors which influence assay performance.

## 11.5  Addresses

### Reference standards

PhEur Chemical Reference Substance, The Secretariat of the European Pharmacopoeia Commission, Council of Europe, 6700, Strasbourg, France.

United States Pharmacopoeia Reference Standards, 4630 Montgomery Avenue, Bethesda, Maryland 20014, USA.

World Health Organization International Standards, National Institute for Biological Standards and Control, Blanche Lane, South Mimms, Potters Bar, Hertfordshire, EN6 3QG, United Kingdom.

### Culture supply

American Type Culture Collection, 12301 Parklawn Drive, Rockville, MD 20852, USA.

National Collection of Type Cultures, Central Public Health Laboratory, 61
   Colindale Avenue, London, NW9 5HT, United Kingdom.
National Collection of Industrial and Marine Bacteria, 23 St Machar Drive,
   Aberdeen, AB2 1RY, United Kingdom.

## Equipment

| | |
|---|---|
| Re-usable 30-cm square glass assay plates | Mast Laboratories, Bootle, Merseyside, UK |
| Disposable 25-cm square polystyrene assay plates | Life Sciences, Paisley, UK |
| Diluters | V.A. Howe & Co, Banbury Oxon., UK<br>Tecan UK, Reading, Berks, UK |
| Plate reader | Scientific and Technical Supplies, Newmarket, Suffolk, UK |
| Stomacher | Merck Laboratory Supplies, Magna Park, Lutterworth, Leics., UK |

## References

ANDREWS, J.M. (1999) Microbiological assays. In: REEVES, D.S., WISE, R.,
   ANDREWS, J.M. and WHITE, L.O. (eds), *Clinical Antimicrobial Assays*. British
   Society for Antimicrobial Chemotherapy, Oxford University Press, pp. 35–44.
*British Pharmacopoeia* (1999), Vol. 2. The Stationery Office, London, pp. A205–10.
*European Pharmacopoeia* (1997), 3rd edn. Council of Europe, Strasbourg, pp. 104–11.
HEWITT, W. and VINCENT, S. (1989) *Theory and Application of Microbiological Assay*.
   Academic Press, London.
*United States Pharmacopoeia XXIII* (1995) United States Pharmacopeial Convention Inc.,
   pp. 1690–6.
WARDLAW, A.C. (1985) *Practical Statistics for Experimental Biologists*. John Wiley,
   Chichester, pp. 229–36.

## Further reading

REEVES, D.S., WISE, R., ANDREWS, J.M. and WHITE, L.O. (1999) *Clinical Antimicrobial
   Assays*. British Society for Antimicrobial Chemotherapy, Oxford University Press.

# 12

# Disinfection and Cleansing

NORMAN HODGES[1] AND ROSAMUND M. BAIRD[2]

[1]*School of Pharmacy and Biomolecular Sciences, Brighton University, BN2 4GJ, United Kingdom*
[2]*Department of Pharmacy and Pharmacology, University of Bath, Claverton Down, Bath BA2 7AY, United Kingdom*

## 12.1   General considerations and terminology

Effective cleaning and disinfection of manufacturing facilities are crucial to the achievement and maintenance of the high-quality standards required of medicines and medical devices. The range of operations in pharmaceutical manufacturing is so large that it is not possible to specify a cleaning and disinfection protocol which is universally applicable, and so the strategy adopted for this chapter is one of identifying the factors to be considered and the procedures likely to be applicable in the most critical scenario, i.e. the aseptic manufacture of a medicine or device from individual sterile components, and then identifying the less rigorous practices which might be acceptable in other circumstances.

In many industries cleaning of the premises is an operation which is perceived to have minimal impact on the quality of the manufactured product, and so assumes a low level of importance for both management and cleaners. This situation is unacceptable in the pharmaceutical industry because inadequate cleaning or disinfection not only affects product quality but also inevitably leads to conflict with the regulatory authorities, and possibly to large financial penalties from resulting production losses. For these reasons it is important that the cleaning of manufacturing and testing facilities is carried out by staff who have adequate training, hence ensuring an understanding of the operation of clean room practices, the factors which predispose to contamination of products, and the factors influencing microbial growth and survival.

Rarely do management and supervisory staff responsible for clean room operations find themselves in the fortunate position of having the opportunity to design and commission a new clean room; more frequently, they become responsible for one that is already operational. If, however, a new facility is to be built, the ease with which it can be cleaned and, indeed, the whole cleaning strategy are factors which merit careful consideration. The design of the room is outside the scope of this chapter, but is considered in detail by White (1990). The cleaning strategy, however, may centre upon the choice between manual cleaning and the installation of a (semi-) automatic clean-in-place (CIP) or sterilize-in-place (SIP) facility. These are discussed further in section 12.3, and it is sufficient at this point to emphasize that while CIP/SIP systems clearly represent a higher initial capital cost, this may be offset by the elimination of labour costs resulting from manual cleaning. CIP/SIP systems also minimize the validation problems which result from the potential lack of reproducibility inherent in a manual process. Regardless of the method chosen, the success of any cleaning system is dependent upon operator care and thoroughness. Whichever option is selected the cleaning and disinfection protocol should:

- achieve the required levels of cleanliness in respect of particulate contamination, chemical residues and micro-organisms;

- represent minimal risk to the operator (avoidance of hazardous or toxic cleaning agents);

- not itself leave chemical residues;

- not cause damage or deterioration of work surfaces or equipment;

- avoid the creation of circumstances which predispose to multiplication of micro-organisms already in the room; and

- not add to the bioburden within the room.

Regardless of the protocol involved, it is assumed that there is an effective vermin control policy in place to exclude insects, rodents and birds.

The cleaning and disinfection procedure should be fully documented with appropriate standard operating procedures (SOPs), records and validation data. The validation aspect is one which may attract particular regulatory attention, and there are guidance notes available from the United States Food and Drug Administration (FDA, 1993) and in the Rules and Guidance for Pharmaceutical Manufacturers and Distributors (Orange Guide) (Anonymous, 1997). The terms used in these documents merit consideration because they are not always precisely defined. 'Cleaning' is simply the removal of undesirable substances from surfaces; this renders the surface visually clean and usually facilitates its subsequent (or simultaneous) disinfection. The term 'sanitization' is also commonly employed and is normally understood to mean a single process combining both cleaning and disinfection in one operation. Disinfection, which is the inactivation (killing) or removal of micro-organisms from the surface of inanimate objects, should be distinguished from sterilization which is an absolute term meaning the *complete* removal or inactivation (killing) of all living organisms. A disinfected surface, therefore, is not necessarily sterile; many species of bacterial spores, some fungal spores and viruses may survive exposure to disinfectants.

## 12.2 Implementation of a cleaning and contamination control programme

There are several distinct steps in the implementation of a cleaning and contamination control programme: selection of protective clothing and equipment; selection of cleaning agents and disinfectants; application of those agents; and validation of the process.

## 12.2.1 *Protective clothing and equipment*

It is a fundamental principle of cleaning that the process itself should not introduce one form of undesirable material into the clean room while at the same time removing the 'target' substance(s). Precautions to avoid the introduction of dirt and micro-organisms by *cleaning* personnel should, therefore, be exactly the same as those adopted by the normal operators; similar training, personal hygiene and clothing considerations will therefore be applicable.

Clean rooms are supplied with air which has passed through a high-efficiency particulate air (HEPA) filter, so the air in an unmanned clean room is of very high quality with an aero-microbial contamination level normally well below 1 colony forming unit (c.f.u.) $m^{-3}$ (White, 1990). Most particulate contamination which arises in a manned room originates from skin scales shed by operators; clean room clothing is designed to minimize the escape of such scales. The clothing requirements for each grade of clean area, based upon the Rules and Guidance for Pharmaceutical Manufacturers and Distributors (Anonymous, 1997), are listed in Table 12.1.

The equipment which is used for cleaning should, itself, be cleaned and disinfected or autoclaved, and must be designed to minimize its potential to attract and retain dirt

**Table 12.1** Clothing requirements for personnel in different grades of clean area

| Room grade | Clothing requirements |
|---|---|
| A/B (Class 100)* | Headgear should totally enclose hair, and, where relevant, beard and moustache; it should be tucked into the neck of the suit<br>Sterilized face masks, non-powdered gloves, sterilized or disinfected footwear should be worn<br>Trouser-bottoms should be tucked into footwear and sleeves tucked into gloves<br>Clothing should shed virtually no fibres or particulate matter, and should retain particles shed by the body |
| C (Class 10 000)* | Hair, beard and moustache should be covered<br>A single or two-piece trouser suit, gathered at the wrists with a high neck<br>Appropriate shoes or overshoes<br>Clothing should shed virtually no fibres or particulate matter |
| D (Class 100 000)* | Hair, beard and moustache should be covered<br>General protective suit<br>Appropriate shoes or overshoes<br>Appropriate measures should be taken to avoid contamination from outside the clean area. |

*US Federal Standard 209E designations.
Adapted from Rules and Guidance for Pharmaceutical Manufacturers and Distributors (Orange Guide) (Anonymous, 1997).

**Table 12.2**  Preferred design features of cleaning equipment

| Equipment | Design feature |
|---|---|
| Cleaning cloths | Disposable and must not shed fibres |
| Mop heads | If polyurethane foam, should not be brittle with the potential to shed particles (use as a disposable item) |
| Buckets | Stainless steel to provide a smooth scratch resistant surface which can be autoclaved |
| Mop handles | Stainless steel or plastic, not wood |
| Vacuum cleaners | Fitted with absolute filters |
| Miscellaneous | Items of solid construction rather than laminated (which might peel apart) with flat surfaces, preferably of stainless steel. |

and micro-organisms when not in use. Table 12.2 provides examples of preferred design features of cleaning equipment. Generally, the preference is for items which have flat, non-scratchable, easily cleaned and sterilized surfaces which are not liable to disintegrate, shed particles or retain moisture after use. Brushes have no place in the clean room; not only are they difficult to clean and disinfect, but they also create dust clouds during use.

### 12.2.2  *Selection of cleaning agents and disinfectants*

Selection of cleaning agents (detergents) and disinfectants will be influenced by consideration of: the range and composition of the materials/equipment to be cleaned and the organic material to be removed; the number and types of micro-organisms expected to be present before cleaning and the maximum residual levels acceptable on completion; whether cleaning and disinfection will be separate operations or a single-step procedure (sanitization); and whether a manual or (semi-) automatic CIP operation is selected. Additionally, compatibility between cleaning materials and the cleaning agent/disinfectant should be established; in particular, plastic sponge mops may inactivate phenolic disinfectants (Maurer, 1985). A summary of materials which may inactivate various disinfectant groups is provided in Table 12.3.

Surfaces and equipment in pharmaceutical manufacturing and testing facilities may be contaminated with a wide range of soiling materials, including product residues from a previous batch of manufactured product, lubricants, dust and micro-organisms. These may vary in their water solubility and the extent to which they are dried onto the surface. Cleaning agents need to be selected taking account of these factors, as well as the chemical compatibility of the materials and work surfaces to be cleaned by the detergent which may be formulated as acidic, caustic or neutral. Health and safety issues relating to the detergent and the ease with which it can be removed by rinsing are also factors that may influence the choice; obtaining a residue-free surface is difficult, but it can be achieved with 60–80% alcohol preparations.

Similar considerations govern the selection of a disinfectant, but here, additional factors arise relating to the type and concentration of micro-organism to be killed, the extent to which the disinfectant activity is reduced by organic soiling (more important in a sanitization process where the disinfectant is not applied to an already cleaned

**Table 12.3**  Inactivation of chemical disinfectants

| Disinfectant groups | Inactivation by: | | | |
|---|---|---|---|---|
| | Hard water | Organic material | Other natural materials | Man-made materials |
| Alcohols: ethyl, isopropyl | Slight | Slight | Slight | Slight |
| Formaldehyde | Slight | Slight | Slight | Slight |
| Amphoterics | Serious | Serious | Serious | Serious |
| Diguanides: chlorhexidine | Serious | Serious | Serious | Serious |
| Halogens: chlorine compounds | | | | |
|   1. Hypochlorites | Slight | Serious | Slight | Slight |
|   2. Chloramines | Slight | Serious | Slight | Slight |
| Halogens: iodine compounds | | | | |
|   1. Iodophors | Slight | Serious | Slight | Slight |
|   2. Povidone iodine | Slight | Serious | Slight | Slight |
| Phenolics | | | | |
|   1. Chloroxylenol | Serious | Serious | Moderate | Moderate |
|   2. Clear soluble | Slight | Slight | Moderate | Moderate |
|   3. White fluids | Slight | Slight | Moderate | Moderate |
| Pine fluids | Serious | Serious | NT | NT |
| Quaternary ammonium compounds (QACs) | | | | |
|   1. Cetrimide | Serious | Serious | Serious | Serious |
|   2. Benzalkonium chloride | Serious | Serious | Serious | Serious |
| QACs and diguanides | | | | |
|   Cetrimide/chlorhexidine | Serious | Serious | Serious | Serious |

NT = not tested.
Other natural materials include cork, wood, cellulose sponge, cotton, paper.
Man-made materials were nylon, polyethylene, polystyrene, polyvinyl chloride (PVC), polypropylene, polyvinyl acetate.
Adapted from Maurer (1985).

surface), and the heat stability of the disinfectant (relevant if elevated temperatures are used in a CIP process). The prevalent organisms in the manufacturing area will be identified during the environmental monitoring programme (see Chapter 8) and their source, e.g. dust or skin scales, will largely determine the species present. In an area which is not supplied with HEPA-filtered air, the numbers of dust-borne organisms are likely to be significant and these will include both bacterial and fungal spores which tend to survive well in the dry state. In a clean room where most micro-organisms originate from the skin of operators, Gram-positive cocci and corynebacteria are more likely to predominate. In terms of antimicrobial spectrum, an ideal disinfectant is one which is active against a wide range of micro-organisms likely to be encountered (Gram-positive and Gram-negative bacteria, yeasts, moulds, viruses and spores), but the agents which possess the broadest activity, e.g. aldehydes and halogens, tend to be the ones exhibiting the greatest potential for chemical incompatibility or human toxicity. The broad chemical classes of disinfectants which might be considered, and

**Table 12.4**   Antimicrobial spectra of different classes of disinfectant

| Disinfectant | Gram-positive bacteria | Gram-negative bacteria | Yeasts and fungi | Viruses | Spores |
|---|---|---|---|---|---|
| Phenolics: | | | | | |
|   coal tar | Good | Good | Moderate | Limited | None |
|   synthetic | Moderate | Moderate | Limited | None | None |
| Halogens: | | | | | |
|   chlorine | Good | Good | Good | Good | Limited |
|   iodine | | | | | |
| Formaldehyde | Good | Good | Good | Good | Moderate |
| Quaternary ammonium compounds | Good | Moderate | Good | Limited | None |
| Amphoterics | Good | Good | Good | Limited | None |
| Alcohols 60–70% | Good | Good | Moderate | Limited | None |
| Biguanides | Good | Good | Moderate | Limited | None |
| Peroxygen compounds | Good | Good | Good | Good | Moderate |

their activity against the major groups of micro-organisms, are listed in Table 12.4. The mode of action, advantages and disadvantages of these different classes are summarized in section 12.5.

### 12.2.3   *Cleaning practices and application methods*

When cleaning or sanitizing a clean room it is necessary to ensure that all cleaning items and liquids which enter the room do not introduce dirt or microbial contamination. Thus, all items, including disinfectant containers, which enter the room should be disinfected on the outside, and cleaning or disinfectant solutions should be prepared with fresh purified or sterile water to minimize the bioburden. Solutions introduced into a Class A or B room should be sterile and made with Water for Injections. Since resistant bacteria may well grow during prolonged exposure to dilute disinfectant solutions, it is good practice only to keep concentrated disinfectant in a clean room and only to prepare a dilute working solution as required; 'topping-up' is unacceptable. Such solutions are prepared by addition of the concentrate to a pre-measured volume of water (rather than vice versa), and this is passed through a 0.2 μm pore size membrane into a clean, sterile container.

It is essential that staff who write SOPs for cleaning or sanitization and the cleaning personnel themselves are aware of the factors which might predispose to bacterial survival and growth. Emphasis should be placed upon the avoidance of *unnecessarily* large volumes of aqueous solutions because these may take longer to dry and so present an opportunity for microbial multiplication. There should be an awareness that while micro-organisms cannot grow without water, they are not necessarily killed by drying; bacterial spores readily survive both drying and exposure to alcoholic solutions.

Manual cleaning procedures may be based upon either a one-bucket or a two-bucket system. In a one-bucket system a germicidal cleaner (sanitizer) is used, the liquid being applied and then removed by a mop or vacuum, without a rinsing step. In a two-bucket system the cleaning or sanitizing solution is contained in both buckets, and the mop or cloth is soaked with liquid from the first bucket and rinsed with liquid from the second. This procedure minimizes the dispersal of dirt which may result if just a single bucket is used and its contents become progressively dirtier. If cleaning and disinfection are two separate operations, it is necessary to have a rinse step following each. Cleaning normally proceeds from high level in the room downwards, from the areas nearest to the HEPA filters to those more distant, and from the area furthest from the door. It is essential that the filters are not wetted with the cleaning agent and that consideration is given to electrical safety; electrical sockets should have leak-proof gaskets, but the power should, nevertheless, be switched off until the room is dry. Excess water may be conveniently removed by vacuum, and this is particularly important where water enters crevices. Residues of cleaning or sanitizing solutions may be removed using cloths soaked with pre-filtered alcohol. At the end of the cleaning procedure all waste materials, including cleaning agents and water, should be removed from the area and all cleaning equipment should be immediately cleaned.

Alternative methods of cleaning may involve 'fogging', whereby a fine disinfectant mist is generated by atomizer. Fogging systems can be built into production units by installing fogging nozzles supplied with pre-diluted disinfectant, the fog being generated by compressed air. Several fogging systems are available, and the manufacturer's advice should be sought for the arrangement best suited to the unit in question. With regard to suitable fogging regimens, generally 1 litre of pre-diluted fogging disinfectant is applied to $100\,m^3$ of air space. The strength of disinfectant is that recommended by the manufacturer; droplet size can be adjusted if necessary. The location and number of fogging systems will depend upon the size and shape of the room; frequently they are located at bench height, facing into the centre of the area. It may be necessary to protect air filters during the fogging operation in order to prevent them becoming saturated and non-functional. A number of Health and Safety considerations must be addressed. As fogging agents may be surface active agents or harmful in their own right, personnel must be banned from entering the facility while fogging is in progress, and respiratory protection must always be available. The dwell time must also be established, as must be the period after which it is safe for operators to enter the fogged area, typically 12–24 h.

### 12.2.4 *Cleaning, disinfection and sterilization of isolators*

Guidance on the cleaning, disinfection and sterilization of isolators is sparse in the literature, and manufacturers of isolators are often not forthcoming in offering assistance. As far as cleaning and disinfection is concerned, the purpose is to reduce microbial counts inside the isolator to an appropriate and acceptable level. This entails first removing any transient contaminants by cleaning the internal surfaces of the isolator and the external surfaces of resident equipment with a low-residue detergent, and then disinfecting using a sporicidal or non-sporicidal agent (e.g. halogen compounds, hydrogen peroxide, peracetic acid, formaldehyde and glutaraldehyde or alcohols, phenols, quaternary ammonium compounds and chlorhexidine, respectively). Surfaces of components used in isolators may contain transient contaminants, and these are best removed by application of alcoholic preparations. Clearly, any disinfectant used in an isolator must be of an appropriate microbiological quality; aseptic filtration may be required to achieve this. Cleaning and disinfection procedures should be clearly documented, precisely followed, and adequately recorded.

The sterilization of isolators involves the introduction of an appropriate gaseous agent; this reduces the likelihood of microbial contaminants in the area or on surfaces to a predetermined and acceptable level, thereby increasing the sterility assurance of products made within the isolator. The sterilants most frequently used are vapours of peracetic acid or hydrogen peroxide; ethylene oxide or formaldehyde vapour have also been used. The choice will be determined by both process- and equipment-related factors. Concentration, humidity, air flow patterns and contact time must be precisely known and controlled. Sterilization failure can occur through inefficient contact with surfaces; even minimal occlusion, caused for example, by a glove resting on a work surface, may result in reduced lethality. Isolator loading patterns must therefore be validated during process development and rigorously followed during operational procedures. Microbiological validation of the process may be performed with biological indicators (BIs); spores of *Bacillus stearothermophilus* are recommended in the *European Pharmacopoeia* (EP) (1997) for use with hydrogen peroxide- and peracetic acid-based processes. Once the target lethality is achieved, the contact time and sterilant concentration can be selected to include an acceptable safety margin, making due allowance for sterilant compatibility with equipment and components, and also the need to purge residues before aseptic processing can begin. Contact time should be as short as possible consistent with effective and reliable sterilization, while purge time should be as long as possible. Having established processing conditions, the cycle loading pattern can then be validated using BIs in worst case positions in replicate cycles. Positive controls should be included and recovery conditions verified. Using this information generated from a known microbial population, the lethality of the gaseous sterilizing process to the natural bioburden can then be estimated. Thereafter, routine cycle monitoring should be properly documented and controlled. For further information, the reader is referred to Lee and Midcalf (1994).

### 12.2.5 *Standard operating procedures*

Many of the parameters which should be specified in a SOP for cleaning and disinfection relate to the cleaning agents themselves and the mode of application. However, there are several other details commonly expected by the regulatory authorities; a more complete list would include:

1  Designation of persons responsible for cleaning, and details of their supervision.
2  Frequency of cleaning and the maximum permissible time that may elapse between completion of processing and cleaning.
3  Details of any differences in cleaning procedures applied between different batches of the same product compared with different products.
4  Specifications for protective clothing and equipment (including design and permitted re-use, if any).

5 Chemicals to be used: their concentration, contact time, filtration conditions, rinsing and rotation if any (to minimize selection of resistant organisms).

6 Method(s) for applying and removing cleaning agents.

7 Method and extent of disassembly of equipment to facilitate cleaning.

8 Procedures for removing or obliterating previous batch identification.

9 Procedures for protecting clean equipment from contamination prior to use.

With regard to the rotation of disinfectants, the Orange Guide (Anonymous, 1997) simply states that where disinfectants are used, more than one type should be employed. In practice, this is interpreted as rotation on a regular basis, ranging from periods of 2 weeks and perhaps up to 6 weeks, using disinfectants with a different mode of action. Thus, amphoterics, biguanides or quaternary ammonium compounds (QACs) would be classified in the first rotation group and phenolics, hypochlorites, aldehydes, chlorine dioxide or peracetic acid in the second rotation group. Some companies might use an uneven rotation, reserving the use of the more hazardous compound only for resistant organisms. Many microbiologists are, however, of the opinion that rotation of disinfectants creates both operational and technical difficulties in practice. In particular, dosing equipment requires thorough rinsing to remove all traces of the previous chemical; neutralization of disinfectant activity may otherwise occur. With the development of more sophisticated disinfectant formulations, the need for rotation of disinfectants may be reduced, provided that cleaning and disinfection procedures are shown to be effective and that disinfectant concentration and biocidal performance is regularly monitored.

## 12.3 Clean-in-place (CIP) and sterilization-in-place (SIP) facilities

CIP/SIP facilities are semi-automatic or fully automatic, and are designed to achieve effective cleaning or sterilization by fill and soak/agitation, solvent refluxing or spraying, without the need for equipment disassembly. CIP systems were developed principally in the dairy and brewing industries from which much of the authoritative literature derives (Romney, 1990); their adoption in the pharmaceutical industry has been somewhat slower, due, perhaps, to the exacting standards which are required and the difficulties which may arise in validation (see below). CIP systems normally employ cleaning or sanitizing solutions, often at elevated temperatures which are pumped under pressure from mixing tanks and the solutions may be recirculated during single use or re-used for multiple cleaning cycles. Such a system, therefore, incorporates tanks, pipework, pumps, valves, spray devices and control and recording equipment, and is so complex that it is normally incorporated in the manufacturing facility at the design stage rather than added later. This complexity might also present additional problems at the validation stage because of inaccessibility of parts of the system. A well-designed CIP system may achieve more effective and reproducible cleaning than a manual operation, but a poorly designed system can, itself, pose problems, e.g. additional pipework and electrical supply cables within the clean room may create cleaning difficulties by promoting the entrapment of dust; unsuitable plumbing may lead to 'deadlegs' in the CIP system and stagnant liquid may act as a breeding ground for micro-organisms. The design of CIP systems is discussed in detail by Seiberling (1987), and further information can be obtained from the CTFA

Microbiology Guidelines (Anonymous, 1993) and Clegg and Perry (1996). CIP programmes for both piping systems and spray cleaning may incorporate: (i) a pre-rinse with cold Water for Injections; (ii) recirculation of the cleaning or sanitizing solution for the required time at a temperature between 55 and 80°C; (iii) a post-wash recirculated rinse which may incorporate dilute acid to neutralize alkaline detergent films; and (iv) a final rinse with Water for Injections.

SIP utilizes specialized equipment for supplying steam, evacuating condensate and air and involves the control, monitoring and recording of cycle parameters and conditions. As with CIP systems, the design of any SIP system is crucial, requiring good engineering practices at the design phase. The validation programme should be designed to qualify both the equipment and the sterilization cycle; it should provide assurance that a cycle run under conditions and parameters established during validation will result in equipment which is both sterilized and suitable for further process use. During validation the system is challenged and must be shown not only to operate and function consistently, but also to be capable of providing uniformly reproducible results. The validation protocol should include the following: an objective defining the qualification requirements of the system and the acceptance criteria; a detailed description of the system; clearly defined responsibilities of those involved in SIP protocols (including set-up, operations, sampling, monitoring, testing, reporting data review and document approval); descriptions for installation qualification (IQ), operational qualification (OQ) and performance qualification (PQ); and detailed acceptance criteria for validation of IQ, OQ and PQ. During performance qualification it must be shown that lethal conditions are achieved; pockets of trapped air can easily accumulate and present problems. Biological indicators, usually *Bacillus stearothermophilus* spores, can be used to confirm that suitable sterilization conditions exist within the system.

For a more detailed discussion of SIP and CIP validation, the reader is referred to Baseman (1992).

### 12.3.1  *Application to isolators*

The principles for cleaning isolators are based upon the classic principles employed in cleaning aseptic areas. The CIP cycle may be based on a manual or semi-automatic system, and should result in effective cleaning of the processing area and all exposed surfaces. SIP then follows by introducing the sterilizing agent in the vapour or gaseous phase into the isolator either through the air handling system or by means of nozzles placed inside the isolator. A suitable generator should control air flow, temperature, pressure and relative humidity. Hydrogen peroxide in the vapour or gaseous phase is commonly used for sterilization. Peracetic acid or formalin may also be used; owing to their corrosive nature and the toxicity of formaldehyde, stringent handling requirements must be observed.

### 12.4  **Validation**

Validation is the generation and collection of data to prove that the cleaning and disinfection process consistently achieves the accepted performance criteria. Guidance on the validation process is offered by the FDA (1993) and by Anonymous (2000);

a collation of information is available on a subscription basis on the web site www.ihshealth.com, and free of charge on the FDA web site www.fda.gov. When planning a validation study, the following should be considered:

1  What data and documents are to be collected?

2  What analytical methods are to be used to determine the concentration of chemicals, chemical residues, micro-organisms and, possibly, endotoxins?

3  What sampling procedures are to be used?

4  What acceptance criteria shall be adopted for cleanliness and disinfection?

### 12.4.1  *Data and document collection*

Clearly, there will be a requirement for well-defined analytical methods and for data on the levels of analytes before and after the cleaning process. However, there are several other items which are less readily anticipated, and a more comprehensive list may include the following:

1  A statement of the principles of the cleaning process. Generally, these should address such questions as whether the process requires scrubbing by hand or simply solvent washing, and how variable are manual cleaning processes from batch to batch and product to product.

2  Data sheets on cleaning and disinfectant products, including information on health and safety aspects, corrosion and chemical compatibility.

3  Supplier audits to ensure quality of cleaning and disinfecting agents.

4  Efficacy data for sanitizing or disinfecting solutions against typical micro-organisms isolated from the manufacturing environment.

5  Identification of the surfaces for which cleaning validation is required (primarily, but not exclusively, those coming into contact with product).

6  Identification of the situations in which validation studies are 'bracketed', i.e. where a single validation study is undertaken on a product, process or item of equipment which is selected as being representative of several similar items.

7  Sampling methods, including the rationale for their selection.

8  Analytical methods, including limits of detection and limits of quantitation.

9  Results of analyses following at least three successful applications of the cleaning procedure.

10  Identification of the frequency or circumstances under which re-validation is undertaken.

11  Identification of the persons responsible for performing and approving the validation study.

12  In process monitoring procedures where relevant, e.g. conductivity monitoring of rinse solutions.

13  Records of routine cleaning operations.

14  A final validation report approved by management which states whether or not the cleaning process has been validated successfully.

## 12.4.2  *Analytical methods*

It is necessary to develop analytical methods which can be shown reliably to detect and, where necessary, to quantify the levels of analytes which are relevant to the adopted criteria of cleanliness or successful disinfection. Thus, conventional chemical methods, e.g. HPLC or total organic carbon determinations, may be combined with microbiological procedures, e.g. total viable counts (or ATP determinations as a substitute) or bacterial endotoxin tests, and, possibly, radio-immunoassay for specific enzymes. The activity of disinfectants against environmental isolates of microorganisms can be assessed using EN 1276 (European suspension test) (British Standard 1997) for example, or by one of the tests for surface disinfection described by Reybrouck (1999). The analytical and sampling methods need, themselves, to be validated; it must be possible to show that contaminants can be recovered from the surface of equipment, and consistently so.

## 12.4.3  *Sampling methods*

Samples for a validation study should be taken according to a written sampling plan. Essentially, there are two methods of sampling which are considered acceptable: direct surface sampling (swabbing) which is regarded as the most desirable (FDA, 1993); or rinse solutions. These two methods are often used simultaneously. Swabbing is a relatively straightforward operation, although it is necessary to conduct a preliminary check to confirm that the ease of detection of the analyte under consideration is not influenced by the nature of the swab material and the solvent selected; for example, adhesive used in swabs has been found to interfere with sample analysis. An advantage of direct sampling is that dried on or insoluble materials can be physically removed for analysis, and it may be possible to quantify the contaminant in terms of amount per unit area. Rinse samples afford the benefits that they permit sampling of larger surface areas and equipment parts which may be inaccessible using a swab, but again, preliminary testing must confirm that the contaminant in question is sufficiently soluble in the solvent for there to be a realistic expectation of its recovery by rinsing.

The FDA (1993) indicate that the strategy of processing a placebo batch of product followed by application of the cleaning procedure may elicit useful information in terms of residual contamination levels from the pre-placebo batch, but the value of the procedure may be limited by dilution of the contaminant, and further problems exist in ensuring that any contaminant is uniformly distributed throughout the placebo batch. Both the FDA (1993) and Anonymous (2000) emphasize the unacceptability of the 'test until clean' concept whereby the sequence of testing, sampling and cleaning is repeated until an acceptable residue limit is attained. This practice is not considered an appropriate substitute for conventional validation methods.

## 12.4.4  *Acceptance criteria*

In view of the wide diversity of equipment and manufacturing facilities in the industry it is not possible to specify universally applicable criteria for successful cleaning. The regulatory view is that each individual manufacturer should establish residue limits which are practical, achievable and realistic in the context of the sensitivity of analytical methods. The only numerical values offered for guidance (FDA, 1993; Anonymous, 2000) are that not more than 0.1% of the normal therapeutic dose of any product will appear in the maximum daily dose of the following product, and no more than 10 p.p.m. of any product will appear in another product.

The value of visual inspection of the cleaned surfaces should not be overlooked. It would certainly be unsatisfactory if chemical or microbiological analyses indicated an effective cleaning or sanitizing process yet visual inspection indicated otherwise, so 'clean by visual inspection' should be a criterion of acceptability.

In a number of cases, the active ingredient of one product should be undetectable in a following product, so the test result simply becomes 'below the minimum detectable limit, therefore acceptable'; this criterion tends to be used particularly for allergenic products, β-lactam antibiotics, steroids and cytotoxics.

## 12.5  Disinfectant monographs

In this section brief monographs are presented, detailing the disinfectant groups in common usage. For each group, the mode of action, advantages and disadvantages are given. For more detailed information, the reader is referred to the manufacturers' literature and to Russell *et al.* (1999).

### Alcohols

*Mode of action:* rapid protein denaturation, cell lysis.
*Advantages:* broad spectrum of activity, no residues left, rapid action.
*Disadvantages:* flammable, not sporicidal, no surfactant properties, expensive, limited activity range (60–80%).

### Aldehydes

*Mode of action:* protein denaturation and cross-linking reactions; reacts with amino groups in vital cell proteins but amino, imino, hydroxyl and sulphydryl groups of all macromolecules are susceptible.
*Advantages:* broad spectrum of activity, including spores, non-corrosive, suitable for aerial disinfection.
*Disadvantages:* toxicity, instability, low soil tolerance.

### Amphoterics

*Mode of action:* cell wall/membrane lysis, disruption of phospholipid molecules in cell membrane, leakage of vital functional chemicals.

*Advantages:* broad spectrum of activity, low toxicity, some surface activity, good stability, easily removed on rinsing.
*Disadvantages:* inactivated by high levels of soil, not sporicidal.


## Biguanides

*Mode of action:* cell wall/membrane lysis, disruption of phospholipid molecules in cell membrane, leakage of vital functional chemicals, coagulation of cytoplasm at high concentrations.
*Advantages:* broad spectrum of activity, low toxicity, non-corrosive.
*Disadvantages:* non-sporicidal, activity limited to specific pH range, less activity against Gram-negative bacteria, activity affected by hard water.


## Chlorine dioxide

*Mode of action:* oxidation of cell proteins, reaction with amino acids, especially those containing sulphydryl groups, similar mode of action to hypochlorites.
*Advantages:* rapid action, broad spectrum of activity, including spores.
*Disadvantages:* corrosive, offensive odour, toxicity, low soil tolerance, requires activation.


## Hypochlorites

*Mode of action:* oxidation of protein functional groups, particularly $-SH$ and $-NH_2$; cell disintegration at high concentrations.
*Advantages:* broad spectrum of activity, effective at low concentrations, cheap, rapid kill.
*Disadvantages:* toxic, corrosive, low stability, inactivated by soil, may contain particulates, react with other chemicals.


## Peracetic acid

*Mode of action:* oxidation of protein functional groups, particularly $-SH$ and $-NH_2$.
*Advantages:* broad spectrum of activity, including spores, non-hazardous decomposition products, rapid action, some soil tolerance.
*Disadvantages:* instability, pungent smell, corrosive to soft metals.


## Phenolics

*Mode of action:* cell lysis, cell membrane leakage, coagulation of cytoplasm at high concentrations.
*Advantages:* broad spectrum of activity, soil-tolerant.

*Disadvantages:* harmful, toxic, damaging to plastics, not sporicidal, strong odour, leave surface residues.

### Quaternary ammonium compounds

*Mode of action:* cell wall/membrane lysis, disruption of phospholipid molecules in cell membrane, leakage of vital functional chemicals, coagulation of cytoplasm at high concentrations.
*Advantages:* broad spectrum of activity, non-hazardous (concentrations < 5%), good stability, non-corrosive.
*Disadvantages:* not sporicidal, inactivated by soil, leave residues, less active against Gram-negative bacteria.

## Acknowledgements

We are much indebted to Richard Emerton (Diversey Lever), Mark Phelps (T.H. Goldschmidt) and Rob Walker (CP Pharmaceuticals) for their generous provision of information upon which this chapter is based.

## References

Anonymous (1993) *CTFA Microbiology Guidelines.* Cosmetic, Toiletry and Fragrance Association US, Washington, DC.
Anonymous (1997) Rules and guidance for pharmaceutical manufacturers and distributors. The Stationery Office, London.
Anonymous (2000) Validation master plan, design qualification, installation and operational qualification, non-sterile process validation, cleaning validation. Draft Annex 15 to the Eu Guide to Good Manufacturing Practice. Volume IV, In press.
BASEMAN, H.J. (1992) SIP/CIP validation. *Pharmaceutical Engineering,* **12**, 37–46.
British Standard EN 1276 (1997) *Chemical disinfectants and antiseptics – quantitative suspension test for the evaluation of bactericidal activity of chemical disinfectants and antiseptics used in food, industrial, domestic and institutional areas – test method and requirements.* British Standards Institute, London.
CLEGG, A. and PERRY, B.F. (1996) Control of microbial contamination during manufacture. In: BAIRD, R.M. and BLOOMFIELD, S.F. (eds), *Microbial Quality Assurance in Cosmetics, Toiletries and Non-sterile Pharmaceuticals.* Taylor & Francis, London pp. 49–66.
*European Pharmacopoeia* (1997) 3rd edn. Council of Europe, Strasbourg, p. 286.
Federal Standard 209E (1992) *Cleanroom and workstation requirements, controlled environment.* Washington, DC.
Food and Drug Administration (1993) *Guide to the inspections of validation of cleaning processes.* Washington, DC.
LEE, G. and MIDCALF, B. (eds) (1994) *Isolators for Pharmaceutical Applications.* HMSO, London.
MAURER, I.M. (1985) *Hospital Hygiene,* 3rd edn. Edward Arnold, London, pp. 44–5, 67–70.
REYBROUCK, G. (1999) Evaluation of the antibacterial and antifungal activity of disinfectants. In: RUSSELL, A.D., HUGO, W.B. and AYLIFFE, G.A.J. (eds), *Principles and*

*Practice of Disinfection, Preservation and Sterilization*, 3rd edn. Blackwell, London, pp. 124–44.

ROMNEY, A.J.D. (1990) *CIP: Cleaning in Place*, 2nd edn. The Society of Dairy Technology, Huntingdon, Cambridgeshire, UK.

RUSSELL, A.D., HUGO, W.B. and AYLIFFE, G.A.J. (1999) *Principles and Practice of Disinfection, Preservation and Sterilization*, 3rd edn. Blackwell Science, Oxford.

SEIBERLING, D. (1987) Clean-in-place/Sterilize-in-place (CIP/SIP). In: OLSON, W.P. and GROVES, M.J. (eds), *Aseptic Pharmaceutical Manufacturing – Technology for the 1990s*. Interpharm Press, Prairie View, IL, USA, pp. 247–314.

WHITE, P.J.P. (1990) The design of controlled environments. In: DENYER, S.P. and BAIRD, R.M. (eds), *Guide to Microbiological Control in Pharmaceuticals*. Ellis Horwood, Chichester, pp. 87–124.

# 13

# Microbiological Hazard Analysis and Audit: The Practice

MARTIN LUSH

*David Begg Associates, 10 High Market Place, Kirbymoorside, Yorks, YO6 6AX, United Kingdom*

## 13.1 Objectives

All involved in pharmaceutical and medical device manufacture will realize the versatility and robustness of microbes and their ability to cause havoc in products, processes and premises without warning. The skilled auditor can, however, make a pre-emptive strike! By identifying potential contamination sources, and by recognizing the vulnerability of product and process, the auditor is well placed to minimize – if not prevent – poor microbiological control. In order to be effective, the auditor requires some essential skills and a good understanding of: microbial ecology and the essential characteristics of micro-organisms; the production process and environment; the product, its characteristics and use; and how to conduct audits in a timely and efficient manner.

The intention of this chapter is to assist the auditor in applying his or her knowledge of pharmaceutical microbiology and contamination control in an auditing capacity. The objectives are to: demonstrate how to prepare, manage and follow-up audits; emphasize the importance of 'hazard analysis' in the auditing process (auditors can utilize their time most effectively by focusing on specific areas of the process at greatest risk of microbiological contamination); outline the key personal skills needed by the auditor, essential for success; and illustrate how to conduct an audit of a pharmaceutical microbiology laboratory and a manufacturing process from early planning to follow-up.

## 13.2 Planning and management of audits: the key to success

### 13.2.1 Why audit?

The pharmaceutical industry has gone to great lengths to introduce and maintain systems for assuring that products are fit for their intended use. Auditing must be regarded as an extension of the Quality Management System; when carried out effectively, audits provide a unique opportunity to challenge its strengths and weaknesses, and are crucial to on-going contamination control and improvement.

### 13.2.2  *Who should audit?*

Poor audits just happen; good audits are properly planned and managed. This requires a committed team effort. Whenever the scope of an audit includes issues relating to microbial control, the microbiologist must be considered an essential part of the audit team. Although the size and make-up of the audit team will be influenced by the purpose and scope of the audit, the key roles of audit leader and secretary should be defined and discharged to the best qualified person at the earliest opportunity.

### 13.2.3  *Audit leader: responsibilities*

The audit leader is responsible for proposing the purpose and scope of the audit and seeking agreement and confirmation from both the auditee and audit team members. He or she also acts as principal contact with the auditee at each stage of audit planning, management and follow-up. He or she is required to lead the audit in an efficient and professional manner by effectively utilizing the team's skills and resources, by being sensitive to the needs and concerns of the auditee and by managing any wrap-up or feed-back sessions. The audit leader also acts as the principal author of the final report and reviews and monitors any follow-up actions. Depending upon the scope of the audit, the leader is responsible for assembling a team with the required technical background and experience; however, the larger the team, the harder coordination and overall management become. Audit teams in excess of two to three persons are rarely warranted.

### 13.2.4  *Planning the audit*

The key to successful auditing is time management. Adequate time should be allocated not only to planning the audit with the audit team and laboratory manager, but also to preparing for the audit by assembling and reviewing all relevant information. Time scheduled for the audit itself should be sufficient for investigating both planned activities and unforeseen circumstances. For the inexperienced auditor, it is often difficult to predict how much time should be scheduled for planning and preparation activities. This will be determined by the skill and experience of the audit team and the purpose and scope of the audit.

### 13.2.5  *Key elements of the planning process*

Before detailed planning begins, there should be broad agreement on the purpose and scope of the audit and the desired, but realistic, time frame for completion. The auditee should become involved, and feel part of the audit at the earliest opportunity. Since the audit remit will be determined by the company's needs, others should be consulted, even though not directly involved in the audit, since they may highlight some important priorities for the audit team. Where, for example, a microbiology laboratory provides agar plates for monitoring a controlled environment, talking to the production operators involved in monitoring could highlight possible problems that the auditor may wish to investigate further. As soon as the broad remit of the audit is agreed, more focused preparations can begin. Once the audit date is confirmed, planning meetings should also be set.

### 13.2.6 Information gathering

Conducting an audit without adequate background information is like assembling a jigsaw puzzle without first seeing the complete picture. Important background information for an audit of a microbiology laboratory may include: an organization chart of the company and the laboratory; a schematic of the laboratory layout; a list of key tasks and analyses routinely undertaken; confirmation of working rotas (shifts, weekends); a summary of contract arrangements, e.g. sterility testing and media preparation; and confirmation of reference standards complied with (e.g. company-specific policies, pharmacopoeias).

This information can come from a variety of sources, including pre-audit questionnaires or a short visit to the laboratory. In addition to providing valuable information to focus preparation activities, this process also helps to build the all-important relationship between auditor and laboratory team.

### 13.2.7 Administration

As the preparation plans gain momentum, essential administration is needed. A provisional agenda for the audit should be issued for comment and confirmation.

### 13.2.8 The opening meeting

Good audits have one thing in common; they are conducted in a relaxed but business-like environment which begins, in earnest, at the opening meeting when a rapport should be established with the auditee. An agenda for this short meeting should cover: personal introductions of all key people; purpose and scope of the audit; audit plan for the day; likely restrictions that the audit team need to consider (restricted access, safety regulations, specific clothing requirements, etc.); and time for others to comment.

### 13.2.9 During the audit

A systematic and organized approach is required, most easily achieved through the use of an aide-memoire or checklist. Questions should be clearly and concisely phrased and directed at those best qualified to answer them, often the 'coal-face workers'. Clearly, notes taken should be legible, orderly and should accurately reflect what is being said.

### 13.2.10  *Key personal skills*

Auditors with good technical skills and understanding, but lacking the required personal attributes, will never achieve their full potential as auditors. Good auditors share essential attributes in being able to:

- recognize that being audited is uncomfortable for most people. As such, they should take steps to put the auditee at ease and in so doing enhance the quality of information provided.

- listen; they seek first to understand and then be understood.

- appear interested in everything, even if they are not.

- acknowledge good practice.

- value and respect the views of others, even if they do not fully agree. Good auditors learn to suspend judgement and accept that they do not have the monopoly on good ideas.

- allow people to save face. There is no value in continuing to emphasize a particular problem repeatedly. Record the concern and move on.

- recognize that conflict is destructive and must be avoided.

One essential skill of singular importance is that of asking questions. The quality of the answer provided is a direct reflection of the quality of the question asked. Questions should be open, unambiguous and concise, and asked one at a time. Closed (yes/no) questions will not encourage people to talk.

### 13.2.11  *Review meetings*

For audits exceeding one day, it is worth having a review meeting at the close of day with the auditee. This provides an opportunity to discuss the audit's progress, follow-up issues raised during the day, and areas for further review. The audit is not complete until the final 'wrap up' session has taken place. This concluding meeting provides the auditor with the opportunity to: summarize the key points of the audit, both good and bad; seek clarification on any outstanding issues not fully understood; follow-up any specific issue left unresolved during the audit; and thank the auditee for their cooperation and participation. Inexperienced auditors often find this meeting difficult, and may try to avoid it.

### 13.2.12  *Classification of deficiencies*

The auditor should classify his or her findings based upon the significance of the observations made. The following scheme may help.

### *Critical deficiency*

This has a high probability of resulting in patient risk, adverse reaction, injury or death. It could result in product recall because of inadequate potency, purity, identification or safety of the product. In addition, serious violations of licence conditions could also be considered as critical.

## Major deficiency

This has potentially serious implications for product quality and could lead to an 'out-of-specification' situation, or close to the 'edge of failure'. Initial impact may not be associated with immediate patient safety, but long-term implications could eventually affect it. Usability or saleability of product could be impaired.

## Other deficiencies

These cannot be ignored, because they represent non-compliance in Good Manufacturing Practice (GMP) terms (or in terms of the company's own declared methods/procedures), but have a low probability of causing product failure and are unlikely to affect patient safety. Large numbers of these observations may be symptomatic of a more serious situation; hence, other deficiencies cannot be considered in isolation.

### 13.2.13 *Audit report*

An audit is not complete until the final report has been issued. The report represents a permanent record of activities and findings and should be an accurate reflection of the audit. The written report should contain no surprises for the auditee! In writing the report, the author should remember that the report is judged on its quality, clarity and ease of reading – not on its quantity. Every report should have a clear structure, for example: introduction to audit; audit objectives; scope (or focus) of audit; executive summary, including a balanced viewpoint on position statement, company strengths and company weaknesses; matters of concern (classified according to chosen scheme, with explanation of the scheme); recommendations; and finally, acknowledgements.

### 13.2.14 *Follow-up*

This is often where audit and self-inspection programmes fall down: the audit is complete, the report is written, then it is filed and forgotten! A nominated individual should take overall responsibility to ensure that necessary action is progressed as planned. Depending upon the urgency, senior management may need to be aware/ involved/approve the actions proposed. For longer-term projects, a plan is strongly recommended which clearly states: what is required; when it will be progressed (this may be dependent on delivery dates, etc.); who is responsible; and who will have authority to state that a particular action has been successfully completed.

## 13.3 Auditing the microbiology laboratory

The microbiology laboratory plays a vital role in the day-to-day testing, checking and monitoring of all phases of the production process from raw material to end-product sampling. Furthermore, microbiology has a fundamental part to play in: the training and education of the work-force; providing advice on good contamination control practices, including formulation development, process and facility design; problem solving, especially providing expert interpretation of unusual results; risk assessment in terms of interpreting the significance of microbiological data in relation to product and patient risk; and in ensuring that the philosophy of 'prevention rather than cure' pervades the entire operation from beginning to end.

Since so much is dependent upon the skills and expertise of microbiologists, their routine procedures and practices must be regularly reviewed by audit. This becomes critically important when independent contract laboratory services are used.

### 13.3.1  *Overview information*

In prioritizing where the audit team will focus its effort, it is important to have a good overview of laboratory activities, perhaps provided by the pre-audit questionnaire. Such information could include: product range and manufacturing processes; reference standards; summary of activities and key tasks completed; organization and management; schematic of the laboratory; and recent and proposed changes.

### 13.3.2  *Product range and manufacturing*

It is important to establish which products and raw materials are sampled and tested, and the nature of the manufacturing processes which the laboratory supports. A laboratory supporting a complex, manual aseptic operation demands much attention from the audit team. This would not be the case for a non-sterile, dry product operation where the risk of microbiological contamination is appreciably less.

### 13.3.3  *First impressions*

Before beginning the audit in earnest, the laboratory should be quickly viewed for its organization, tidiness and activities. The environment should be well controlled (heating, ventilation, lighting), and all technicians should be suitably dressed. These first impressions matter, and they may influence the direction that the auditor takes.

### 13.3.4  *Organization and management*

The organization structure should be reviewed to clarify who does what.

- Are there sufficient numbers to manage adequately the workload?
- How are the technicians organized? Are individuals or teams solely dedicated to a particular task, or do they regularly rotate?
- Who is 'authorized' to review and approve data, and are they sufficiently experienced and qualified? In their absence, how – and to whom – is this authority delegated?

• Does the laboratory operate during 'unsocial' hours? If so, how are staff organized, data reviewed and reported and decisions made? Is there adequate management and supervisory cover?

### 13.3.5 *Training and expertise*

As with any operation, laboratory staff have various levels of expertise and come from a variety of backgrounds, both biological and non-biological. Microbiologists may be qualified in medical, food or general microbiology. Although there is much similarity between these disciplines, there are also noticeable differences. The auditor must establish that the initial and on-going training is satisfactory for the task in hand, has been 'validated', and is fully documented in individual training records. Training effectiveness can easily be confirmed by asking technicians not just how a particular task is completed, but why. Training that fails to explain the 'why' behind the 'how' rarely has lasting benefit. A key role of laboratory management is to provide advice and on-going support to their production and Quality Assurance (QA) colleagues, which can have a direct impact on product quality. This may take the form of: recommendations for bioburden control (e.g. disinfection procedures); advice on validation issues (e.g. water systems, autoclaves); or guidance in the investigation and risk assessment of high, out-of-specification counts.

For this advice to be of value, the microbiologist must have a broad understanding of the products being tested, and the principles, practice and vulnerabilities of the manufacturing process and facilities. The laboratory manager should be asked how often he or she leaves the laboratory to visit the production areas. A laboratory-based manager with little appreciation of product and process can provide only limited support and advice to their colleagues.

### 13.3.6 *The laboratory*

#### Security of access

Access must be restricted to those authorized to enter. Visitors must comply with the dress discipline and hand-washing practices essential for minimizing the risk of cross-contamination. It should be made very clear that they are entering a 'biohazard area' where they may be exposed to viable bacteria, some of which are opportunist pathogens.

#### Design and layout

Although the workload placed upon the microbiology laboratory may *sometimes* increase significantly, this *may* not *be always* accompanied by reciprocal development and improvement of the laboratory. Laboratories with inadequate space may well compromise routine activities as well as basic safety. Key issues that the auditor should consider include: whether there is adequate and suitably segregated space for all activities; whether all surfaces are cleanable and in a good state of repair; and whether material movements are satisfactory in order to prevent mix-up and cross-contamination. Sample reception and the handling of contaminated materials are clearly important considerations here.

### Good housekeeping

Laboratories that are well managed, organized and well controlled generally have high standards of housekeeping, irrespective of how busy they are. Untidy laboratories are unsafe, and provide the auditor with little assurance that everything is under control. Housekeeping policy and routine cleaning procedures and records should be reviewed.

### 13.3.7  *Method validation*

The auditor must quickly establish the accuracy, robustness, reproducibility and sensitivity of the cultural methods, which still predominate in pharmaceutical microbiology. Since most methods are principally dealing with living microbial systems, the validation of these methods rarely, if ever, fit the conventional validation model. The variability and diversity seen in any biological system presents a major challenge to the validation microbiologist and to the auditor who must judge whether such validation has been adequately performed. These studies usually seek to confirm that methods satisfactorily release the contaminant from the sample and do not inhibit recovery. This usually involves 'spiking' samples with test organisms at low levels [$\leq 10^2$ colony forming units (c.f.u.) ml$^{-1}$ or g$^{-1}$] and quantifying those recovered. As with any validation exercise, the auditor should confirm that: a validation protocol exists that clearly defines test methods and acceptance criteria; a validation report exists confirming test results, and any recommendations for corrective action or routine operation; and test conditions and requirements are complied with routinely.

Specific issues that the auditor should consider include the test organisms used (type strains, i.e. from official culture collections and those from the production environment) and their control and maintenance. The inoculum used to spike the product should also be considered, and whether any antimicrobial activity in the sample must be neutralized or inactivated before testing can begin, as well as what is considered to be a satisfactory level of recovery.

### 13.3.8  *Equipment validation*

In a similar fashion, the auditor needs to confirm that all equipment used has been validated. In the case of an incubator this includes: temperature control and uniformity; routine temperature monitoring; review and approval procedures for monitoring records and charts; temperature alarm set points; calibration of temperature probes; and cleaning schedules and records. Likewise, for isolators this includes: high-efficiency particulate absorption (HEPA) filter integrity test records; chamber and gauntlet integrity tests; calibration of gauges and transducers; alarm functions; inlet air velocities and air change rates; pressure differentials; air flow patterns; environmental monitoring data (viable and non-viable); transfer procedures; and sanitization/sterilization procedures. In the case of autoclaves, validation includes: instrument calibration; chamber pressure; heat-up and cool-down rates; temperature hold; cycle timing; and empty and loaded chamber mapping.

### 13.3.9 *Equipment maintenance and calibration*

For equipment to continue to operate in its validated state, it must be routinely maintained as part of a planned preventative maintenance programme. The auditor should check not only whether such a programme operates, but also whether critical instruments and their calibration requirements have been clearly defined. Maintenance support is often provided by contractors. If this is the case, their activities must be carefully controlled.

### 13.3.10 *Culture media*

Although rapid, non-culture methods are finding applications in pharmaceutical microbiology, most methods still rely upon a more traditional use of a wide range of culture media. These have been very carefully formulated to provide the essential nutrients and conditions, such as available water and pH, needed to give micro-organisms an opportunity to multiply. Only slight changes in the constituents of a medium, arising through poor preparation practices for instance, can dramatically reduce its fertility. 'Over-cooking' of medium during sterilization, over-drying following excessive storage of agar plates, and slight drifts in pH can all have a detrimental effect. Most media formulations and preparation practices have originated from the disciplines of food and medical microbiology where numbers of bacteria in the original sample are generally very high. Minor fluctuations in media quality that would reduce fertility are, therefore, less significant. This is not the case for the pharmaceutical microbiologist whose sample may contain very low numbers; here, even small variations in medium quality can make the difference between a pass and fail result. Many of the media preparation practices used in food and medical microbiology lack the controls needed by the pharmaceutical industry. This is further compounded by the low priority given to the media 'kitchen' in some laboratories. These are often ill-equipped and under-resourced. This area definitely warrants audit attention!

#### Media preparation: general procedures

The auditor should confirm that each stage of media preparation is adequately described and detailed in Standard Operating Procedures (SOPs), providing guidance on the following: receipt of bulk dehydrated media from a position of batch traceability maintenance, certificate of analysis reception and approval, storage conditions (temperature, humidity, light) and good stock rotation (first in, first out); and the manufacturing process. Media are generally reconstituted, mixed to ensure homogeneity, sterilized and dispensed. The manufacturing process includes the following key steps.

#### Glassware preparation and use

Dedicated glassware should be used for media containing inhibitory factors to prevent any inhibitory components contaminating other media. Detergents used in the washing of used or contaminated glassware need to be removed by adequate rinsing.

## Weighing and reconstitution

The calibration status and range of the balance should be checked to confirm that bulk media are accurately weighed out. Purified water used in reconstitution should be of the right microbial and chemical quality. Before sterilization, the reconstituted media are generally heated to 50–60°C with mixing to ensure all ingredients are adequately dissolved. Temperature/time records should confirm that this is properly controlled.

## Media sterilization

Dissolved media are sterilized by moist heat. The auditor should expect to find the autoclave under good control, validated and properly maintained with all the critical parameters (time, temperature, pressure) monitored and reviewed against defined acceptance criteria. Poor control over time and temperature can lead to media being over- or under-sterilized.

## Dispensing

Following sterilization, the molten, sterilized agar is cooled to 45–50°C and dispensed aseptically into sterile plates or bottles, usually via automatic dispensing machines. The auditor will want to know how the media are mixed during dispensing to ensure homogeneity, how the volume is controlled, whether there are dedicated dispensing tubes for particular medium types, and the environmental conditions where dispensing takes place. Some laboratories adopt the practice of allowing bulk, sterilized media to solidify. This is then stored (protected from direct sunlight) and re-melted as required before dispensing. This practice should be adequately controlled and should have no detrimental effect on medium fertility. Common abuses of this practice include multiple meltings and retaining the molten agar at 45–50°C for long periods.

## Labelling and storage

Following dispensing, medium should be labelled with its identity, batch code and expiry date, packed (usually in sealed plastic bags), and stored ready for use. The auditor needs to know whether the expiry date and storage conditions have been validated. Warm temperatures can cause excessive condensation on the surface of agar plates, and storage in a temperature-controlled environment is essential.

## Quality control testing of prepared media

Fertility of the media should have been checked with documented laboratory strains and perhaps 'wild strains' obtained from the production environment. The inoculum level (usually $<100$ c.f.u. ml$^{-1}$) and test method (Miles and Misra being preferred; see Chapter 4) should have been noted, as well as what is considered to be a satisfactory percentage recovery of test organisms. The auditor should also establish when the pH is measured and testing is done. Plates that are fertile within hours of pouring may not be after sustained storage and drying.

## Media contractors

Subcontracting of media preparation is not uncommon, though it is clearly the microbiologist's responsibility to ensure that all preparation activities are properly controlled. Contractors rarely supply media just to the pharmaceutical industry; the bulk of their media may be manufactured for use by food and medical microbiologists, who may have less demanding requirements. Key issues to consider include: segregation of media production (pharmaceutical versus non-pharmaceutical); and the impact of storage, transportation and any 'secondary sterilization' of media on fertility and sterility. Media provided for environmental monitoring of controlled environments are often sterilized by gamma irradiation to allow almost immediate transfer into aseptic processing areas. Some workers have reported such irradiation to produce toxic free radicals which compromise fertility and/or alter the colonial characteristics of isolates.

### 13.3.11  *Good laboratory practice (GLP)*

#### Sample receipt and storage

The bioburden of samples for testing can vary considerably as a result of delays in the sample reaching the laboratory and the conditions to which the sample is exposed during transportation or storage. The quality of any analysis is thus dependent upon the quality of sample. The auditor should, therefore, review the following: justification for sampling plans being used; adequacy of the sampling procedures; suitability of sampling equipment and environment; and, importantly, the skills of the sampler. He should also record how the sample is received and logged in to ensure that traceability is maintained. Sample storage at weekends and the impact that storage time and temperature have on sample quality should also be queried.

#### Procedures

The auditor should review the index of SOPs and use this as a reference throughout the audit. All procedures covered by SOPs, and all the SOPs themselves should be approved, available, complied with, current and regularly up-dated. Reference standards with which the microbiologist complies should be noted [e.g. *European Pharmacopoeia* (EP)/*United States Pharmacopoeia* (USP) and product licence requirements]. It should also be noted how changes in these reference documents are communicated to the microbiologist.

#### Data management, review and approval

Since the end-product of the microbiologist's endeavours is data upon which important decisions are based, the auditor should test the integrity of the documentation trail.

- How are raw data recorded?
- How are test data reviewed and approved?
- How are results reported and within what timescale? The sooner results are reported, the quicker any corrective action can be taken.
- How are test failures reported and reacted to?

- How are trends established from the data trended? Failure to trend important data such as environmental and water monitoring results may prevent the microbiologist from observing adverse trends in performance which in turn delays effective corrective action. The auditor should review the method of trending used, and how trends are presented, interpreted and communicated to those responsible for taking corrective action. The basis and justification of any action and alert limits should also be challenged.

### 13.3.12 *Areas of specific interest to the auditor*

An audit of the microbiology laboratory usually takes place as part of a product-based audit. An audit of a sterile product or a purified water system, for example, will result in the audit team spending considerable time in the microbiology laboratory. The auditor should make full use of this opportunity to review areas of specific microbiological interest, as discussed below.

*Media fill or process simulation trials*

The microbiologist provides valuable resource and expertise in media fills, a key step in the validation of an aseptic process. The auditor should confirm: whether media were fertility tested before and after completion of the media fill; how the media were sterilized and whether sterility was confirmed before use; whether incubation conditions were adequately controlled; the method of inspecting containers for growth during and following incubation; and what were the pass/fail acceptance criteria?

*Sterility testing*

The adequacy and suitability of the sterility testing procedures should be confirmed. Other relevant issues include: the sterility testing environment (clean room versus isolator); the sterility failure and re-test rate; whether all failures have been fully investigated and any subsequent re-test procedures justified; and which reference method has been used (EP, USP) and whether it has been fully validated.

*Environmental monitoring*

The microbiologist plays a central role in the coordination and management of environmental and personnel monitoring of controlled environments. As such, the microbiologist is in an ideal position to explain the monitoring policies, justify the sampling procedures and practices based upon good science and sound judgement, as well as demonstrate how data are managed, analysed and reported. The auditor should confirm that full use is made of this opportunity.

### 13.3.13 *Hazard analysis of critical control points (HACCP)*

In recent years increasing emphasis has been placed on the concept of hazard analysis of critical control points (HACCP) as a method of quality and safety assurance in the pharmaceutical and medical device industry. The concept has been used extensively in the food industry for many years, and is based upon a disciplined and systematic evaluation of the production chain construed as an ecological system in which the hazard organism(s) interact with the accompanying organisms and the processing parameters. Once established, the effect of processing on both product and its accompanying flora can then be used to optimize and monitor those processing steps which have hazard control potential. HACCP, in common with risk assessment and management, should be regarded as a key tool and as such should be systematically used. It should be seen as a living activity with documentation and processes which are regularly reviewed and improved to reflect the constant change which defines any organization. HACCP is not solely concerned with the manufacturing process: it begins with the purchase of raw materials and ends with use of the product by the patient.

Identification of critical control points is a key activity in setting up an HACCP programme. Such points are clearly process-specific, but generally can be defined as those presenting risks of microbiological contamination and/or multiplication during manufacture, distribution, storage and use, and where, or over which, control is required. By definition, therefore, the entire process must be scrutinized from the purchase of raw materials to the use of finished product, and those points which may have a critical impact on microbiological product quality should be identified and quantified. By using a bracketing technique, i.e. analysis of samples before and after a particular process step, the effect of that process on bioburden can be shown. Hopefully, this effect will prove to be beneficial, i.e. the bioburden will decrease, as in a well-thought out and properly implemented cleaning procedure; however, some process steps may conceivably add to the microbial bioburden, and clearly here control is critical. Control measures include the adoption of three lines of defence: limitation of contamination by selecting raw materials of appropriate quality and applying hygienic manufacturing procedures; minimizing any opportunity for growth throughout the entire production, distribution and storage process; and where necessary, including an appropriate sterilizing process to ensure a safe product is produced.

The HACCP approach to management of quality is discussed in more detail in Chapter 3.

## 13.4 Auditing the manufacturing process

### 13.4.1 *Why audit?*

Microbes have a proven ability to contaminate both product and process. Widely distributed in the general environment, often in considerable numbers, they are extremely versatile and can thrive in hostile environments. Not surprisingly, the pharmaceutical industry goes to great expense to minimize the risk of microbiological contamination. As discussed elsewhere, this is achieved through good-quality raw materials and components, control of the production environment, design of facilities and utilities, contamination control procedures and a well-trained workforce with a good understanding of the importance of basic hygiene, basic microbiology and why contamination control measures are so important (see Chapters 8 and 12).

Any audit of the production environment provides an ideal opportunity to challenge the suitability of contamination control practices. Throughout any such audit, the auditor has the following key questions in mind:

- Is there a potential source of microbiological contamination?
- Is the process vulnerable?
- Is there a risk of contamination reaching the product?
- What is the ultimate risk to the safety, quality and efficacy of the product. In other words, is the patient at risk?

Since there is always a source of microbiological contamination, the auditor must make a balanced judgement as to its significance to the process and, ultimately, to the product. As with any audit, time is limited and the auditor is duty-bound to make best use of the time available. In deciding where effort should be concentrated, the following information should be gathered at the earliest opportunity.

### 13.4.2  *Product information*

Here, relevant factors are the microbiological status of the product (sterile or non-sterile), the physical nature of the dosage form (dry, liquid, cream, ointment), and whether the product supports microbial growth.

### 13.4.3  *Process information*

Factors to be considered here include: the raw materials used in the manufacturing process; whether the process is vulnerable to contamination (i.e. open) or whether the product is well protected (i.e. a closed process); whether the product is exposed to recontamination sources and how the contamination risk is controlled; whether there is any storage of intermediate materials; and the total duration of the manufacturing process.

### 13.4.4  *General areas of interest in the building*

#### Walls and ceilings

Moulds are the most commonly encountered microbes on walls and ceilings, particularly when poor ventilation, temperature and relative humidity control lead to high levels of moisture. Contamination may be excessive where damaged surfaces expose the underlying plaster. Surfaces should be smooth, impervious and cleanable; damaged surfaces should be repaired promptly.

#### Floors and drains

Floors should be impervious to water, cleanable and resilient to day-to-day wear and tear. Floors should be laid flat to minimize the risk of excessive surface water (e.g. washing bays) or, ideally should slope towards a drain. The auditor should pay particular attention to joints, seals and floor-to-wall coving to ensure that surface integrity is maintained. Where floor drainage channels are needed, they should be open, shallow, easy to clean, and drain effectively. The auditor must be aware that any 'static' water can act as reservoirs for Gram-negative organisms, particularly *Pseudomonas* species.

### Doors, windows and fittings

These should be flush-fitting whenever possible. Wood readily absorbs moisture and can generate high numbers of moulds; where present, it should be sealed with a high-gloss paint and any surface damage repaired immediately.

### Equipment

The ability of bacteria to attach to surfaces such as stainless steel and plastic and survive should not be under-estimated. Every piece of equipment has its own particular nooks and crannies where microbiological contamination can reside; internal threads and dead-legs cause particular problems (see Chapter 8). Equipment should have smooth continuous surfaces, free from pits; it should be easily dismantled and cleaned. Pumps and valves should be of sanitary design.

### Cleaning of equipment

Cleaned equipment can be readily recontaminated before use (see Chapter 12). The auditor should review: the quality of water used in final rinsing stage; and how equipment is dried and stored to minimize the risk of contamination by *Pseudomonas* and other Gram-negative bacteria.

### Pipelines

Pipelines must be completely drainable to ensure that trapped fluid does not provide a hospitable environment for the growth of bacteria. Internal surfaces should be smooth and polished to minimize pits where microbes may lodge. Joints and welds should be kept to a minimum, since they may provide a protective haven for micro-organisms. Of note, pipelines are often insulated with sealed lagging material. If the protective outer seal is damaged, the exposed lagging may provide a rich source of mould contamination.

### 13.4.5 *Raw materials*

Raw materials pose a major contamination threat to the product and the production environment, and warrant special attention from the auditor. Untreated raw materials of natural origin contain an extensive and varied microbial population, including potentially pathogenic organisms, such as *E. coli* and *Salmonella* species. In cases of excessively high bioburden, pre-treatment may be needed to reduce the bioburden to an acceptable level, using processes such as heat filtration, irradiation, recrystallization from a biocidal solvent or, where compatible, ethylene oxide gas. In contrast, manufacturing processes for synthetic raw materials are usually hostile towards bacteria. As such, their bioburden is considerably lower.

Irrespective of the type of raw material used, the auditor should confirm that the material is provided by an 'approved' supplier. Confidence in the quality of raw materials comes from confidence in the supplier's manufacturing processes and their quality systems, which have been challenged through audit. Likewise, the sampling programme used should be satisfactory, based upon the nature of the raw material (natural/synthetic), the history and performance of the supplier, and the end use of the raw material. Sampling procedures should also be reviewed. Appropriate environmental conditions should be used to reduce the risk of contamination of both sample and bulk. Sampling equipment should be dedicated and clean. Samplers should be properly trained in aseptic techniques (see Chapter 3 for a more detailed discussion). Warehouse storage conditions should also be reviewed: temperature control should be satisfactory; pest control should be effective; and containers should be positioned so that they do not come into contact with damp, cold surfaces such as walls and floors.

### 13.4.6 *Water*

Water is the principal raw material used in the pharmaceutical industry. When reviewing water systems, usually as part of a 'product-based' audit, the auditor must establish quickly an understanding of the system and how it performs. Key facts to know include whether water is used directly in manufacture, and what grades of water are used. Management and operational issues include who owns the system, its complexity (one or multiple plants), its age and any substantial modifications made, and whether the system provides water for facilities that operate shifts or operate through weekends. A schematic of the system should be provided.

### *Microbiological results*

In establishing whether an adequate system of control operates, data from samples taken from user points over the past 6–12 months should be reviewed noting user points sampled, the range of results and underlying trend, whether action and alert limits are visible and appropriate, and whether trend analysis has been applied to improve interpretation and reporting of results. The time of monitoring should also be noted with reference to when the system was sanitized. Even in the event of zero counts, the microbiological sampling practices and test procedures should be challenged.

### *Essential documents*

Validation documents relating to design qualification, installation qualification, operational qualification and performance qualification (DQ, IQ, OQ and PQ) should be complete, available and current for the water system concerned. As with all equipment and operating systems (including that of the air handling system), it is futile to contemplate any microbiological testing programme unless appropriate DQ, IQ, OQ and PQ documentation exists to prove that the system meets all the required specifications. Only when this has been thoroughly reviewed and approved may any microbiological considerations be addressed. In all cases it should be confirmed that protocols were prepared and approved (with all acceptance criteria justified), detailed tasks were completed, results were formally reviewed and approved, and all test acceptance criteria were achieved.

Final reports should have been completed, reviewed and approved highlighting areas of non-compliance together with any justifications and recommendations for corrective action.

In the case of design qualification, the following items require confirmation: specified water requirements (quality and quantity); defined design principles; defined operation and control levels; defined critical instruments; specified system components; comprehensive system drawing; and pre-delivery inspections (welding, degreasing, passivation). Likewise, for installation qualification, the following should be checked: equipment protocols; pipe slopes; absence of dead-legs; material certificates; equipment tagging; wiring check; calibration certificates; and welding certificates and sample welds. Additionally in operational qualification, the following should have been documented: calibration on site; equipment function; set points and alarms; flow rates (maximum/minimum); hydraulic balancing; and sanitization.

Performance qualification should have been spread over an extended period divided into phases. In phase 1, all parameters should have been reviewed and approved at all sample and user points during every day of a 20 to 30-day period. During phase 2, all user points and worst-case locations should have been examined using a reduced sampling scheme over a 2- to 3-month period. The final phase should have investigated similar sample points on a rotational basis for a period of 6–10 months.

Besides these validation documents, the auditor should review the status of other essential documents relating to the water systems. As before, these should be complete, available and in current use. With regard to operation of the system, these should include: sanitization procedures and records for pipework and specific items of equipment (e.g. carbon beds, reverse osmosis membranes); alarms and action to be taken; equipment log books and specific SOPs; and the planned preventative maintenance policy, schedules and records.

In the case of sampling policy for physical, chemical and microbiological parameters, sampling procedures and practices should be reviewed. These should document what is to be done, when and by whom; limits and standards should be defined. Likewise, the management policy should state who has defined responsibilities and how often the system should be reviewed. The role and control of contractors should be clearly documented. Accurate training records should be kept in all cases.

Finally, microbiological procedures to be reviewed include the following: media preparation and fertility testing; media type used; methods used (filtration, plate count, etc.); incubation conditions (temperature/time); identification policy and procedures; action and alert limits; trend analysis procedures; and finally the reporting of data.

### 13.4.7 *Packaging materials*

Cardboard, paperboard and pulpboard, unless sealed or treated, can provide a rich source of contamination, particularly moulds and Gram-positive bacteria, often as resistant spores. If these materials become moist through poor storage, levels of microbiological contamination can increase significantly. Materials such as glass, synthetic rubbers, plastics and laminates have minimal surface microbial counts. However, if stored with limited protection in dusty or damp conditions and packed for transportation in cardboard boxes, often on damp, dirty wooden pallets, they may contain moulds and bacterial spores.

### 13.4.8   *Effective ventilation*

The aerial route of contamination is common and can be significantly reduced by an effective heating, ventilation and air-conditioning system. For non-sterile products, the provision of coarse-filtered, humidity-controlled air can effectively remove dust particles from vulnerable stages of the process. Humidity and temperature control is important, since this not only provides a pleasant working environment, but also reduces the risk of mould contamination.

### 13.4.9   *Cleaning and disinfection*

Although routine sanitization of surfaces is key to controlling environmental contamination, its importance is often over-looked. Since poor sanitization practices can contribute to environmental contamination, this area is certainly worthy of audit time. The sanitization programme and procedure should be reviewed to confirm the frequency and precise method of cleaning and their scientific basis. The cleaning records should confirm procedural compliance (when, where and by whom). The activity of any disinfectant used should be appropriate for the wide range of environmental contaminants likely to be present. It is good practice to rotate disinfectants on a regular basis. Disinfectants only work effectively if made up correctly; the manufacturer's instructions should be followed and fresh disinfectant solutions should be made up before use. Mops, sponges and cloths can provide an ideal environment for rapid and extensive growth of water-borne organisms such as *Pseudomonas* species. They should be washed regularly, dried thoroughly and inspected for signs of wear and damage and replaced if necessary. Inadequately stored and maintained cleaning equipment can be highly efficient vehicles for spreading micro-organisms throughout the environment.

An auditor can quickly gauge how seriously environmental control is taken by the standard of housekeeping. Untidy, cluttered and dirty work areas leave a lasting impression.

### 13.4.10   *Clothing disciplines and personal hygiene*

From the microbiologist's perspective, clothing is worn for one purpose: to protect the product from a major source of contamination, the operator. The auditor should assess the standard of dress; clean clothing correctly worn should be seen at all times, including that worn by contractors and engineers. A high standard of personal hygiene is the single most important means of reducing the potential for product contamination by the operator. People with a pride in their personal appearance should therefore be selected at employment. Pharmaceutical companies should thereafter provide them with training in the importance of personal hygiene and, importantly, provide them with appropriate facilities and clothing to maintain acceptable levels of cleanliness.

### 13.5   Summary

The role of the auditor is immensely rewarding, but not easy. A thorough knowledge of the product and manufacturing process, complemented by a range of good interpersonal skills are essential in making sound judgements throughout the audit. Good judgement comes from experience, and experience comes from making mistakes! Therefore each audit should always be critically reviewed.

# 14

---

# Case Studies and Worked Examples

---

## 14.1 Calculation of viable count and preservative efficacy test data

**NORMAN HODGES**

*School of Pharmacy and Biomolecular Sciences, Brighton University, BN2 4GJ, United Kingdom*

### 14.1.1 Viable counts

These are calculated from the formula:

$$VC = \frac{N \times DF}{V}$$

where:

VC = viable count (in c.f.u. ml$^{-1}$)
$N$ = the number of colonies counted on the plate
DF = the dilution factor, and
$V$ = the volume of sample (ml) added into or onto agar (typically 1.0 ml for pour plates; <1 ml for surface spread; ≪1 ml for Miles and Misra counts)

If, for example, a bacterial suspension had been diluted one million-fold (i.e. a dilution factor of $10^6$ resulting from six successive 10-fold dilutions), 0.2 ml of that dilution spread onto the surface of each of three replicate plates and 26, 31 and 30 colonies (mean 29.0) were counted after incubation, the viable count would be:

$$VC = \frac{29 \times 10^6}{0.2} = 1.45 \times 10^8 \text{ c.f.u. ml}^{-1}$$

### 14.1.2 Preservative efficacy test data

The effect of the preservative is expressed as the extent to which the theoretical (calculated) viable count following inoculation of the product is reduced by the action of the preservative. Typically, sufficient inoculum culture (at a concentration of approx. $10^8$ c.f.u. ml$^{-1}$) is added to the product under test (in its market container) to produce an initial concentration of $10^5$ to $10^6$ c.f.u. ml$^{-1}$. If the product volume is small (say 10–20 ml), this would entail measuring volumes of the order of 0.1 ml or less, which results in relatively large volumetric errors. Thus, for addition to aqueous products with a specific gravity of 1.0 g ml$^{-1}$, such small inocula are better measured by weight.

If, for example, the concentration of a bacterial inoculum was $1.45 \times 10^8$ c.f.u. ml$^{-1}$ (calculated as described above) and 0.1078 g was used to inoculate 22.0 ml containers of topical product, the initial cell concentration immediately after inoculation would be

$$= \frac{VC \times Wt \text{ of inoculum}}{Wt \text{ of product in container}} = \frac{1.45 \times 10^8 \times 0.1078}{22.0} \text{ c.f.u. ml}^{-1}$$

$$= 7.10 \times 10^5 \text{ c.f.u. ml}^{-1}$$

This is usually expressed as a $\log_{10}$ value $= 5.85$

Assuming that samples of the inoculated product were subject to viable counting in the manner described above at time zero and at 2, 7, 14 and 28 days, the data could be tabulated in the following way:

|  | Time zero | 2 days | 7 days | 14 days | 28 days |
|---|---|---|---|---|---|
| Colony counts | 86, 106, 89 | 32, 30, 34 | 76, 95, 108 | 0, 0, 0 | 0, 0, 0 |
| Mean | 93.7 | 32.0 | 93.0 | 0 | 0 |
| Ten-fold dilutions | 3 | 3 | 1 | 1 | 1 |
| Cell concentration in product (c.f.u. ml$^{-1}$) | $4.68 \times 10^5$ | $1.60 \times 10^5$ | $4.65 \times 10^3$ | $<16.6*$ | $<16.6*$ |
| $\log_{10}$ count in the product | 5.67 | 5.20 | 3.67 | $<1.22$ | $<1.22$ |
| $\log_{10}$ reduction | 0.18 | 0.65 | 2.18 | $>4.63$ | $>4.63$ |

*One colony on a single plate would represent a concentration of 16.6 c.f.u. ml$^{-1}$.

The $\log_{10}$ reduction values are simply obtained by subtracting the $\log_{10}$ count in the product from the $\log_{10}$ value of the concentration immediately after inoculation (5.85). The A performance criteria of the EP require a topical product to achieve a 2 log reduction in bacterial count within 48 h and a 3 log reduction in 7 days; since the log reductions in the example above are 0.65 and 2.18 at 48 h and 7 days respectively, the A criteria are not satisfied. The B criteria are a 3 log reduction in bacteria at 14 days and no increase thereafter; the product does achieve this performance target.

It is worth noting that the time zero count ($\log_{10} = 5.67$) is similar to the theoretical inoculum concentration ($\log_{10} = 5.85$). A time zero count which is markedly lower than the theoretical inoculum is not uncommon, and may be due to very rapid antimicrobial activity of the preservative. Time zero counts which are markedly higher than the theoretical value are more difficult to explain and may be due to unsatisfactory counting precision.

## 14.2 Calculation of air change rates

### ROSAMUND M. BAIRD

*School of Pharmacy and Pharmacology, University of Bath, Claverton Down,*
*Bath BA2 7AY, United Kingdom*

A clean room measuring $3.5\,\text{m} \times 4\,\text{m} \times 2.8\,\text{m}$ is supplied with two HEPA filters measuring $0.5\,\text{m} \times 0.6\,\text{m}$. The average air velocity is known to be $0.45\,\text{m s}^{-1}$. The number of air changes per hour can be calculated in the following way:

Air change rate $= Vs/V$ where $Vs$ is the supply air volume and $V$ is the room volume

Room size $= 3.5\,\text{m} \times 4\,\text{m} \times 2.8\,\text{m} = 39.2\,\text{m}^3$

Area of filters $= 2 \times 0.5 \times 0.6\,\text{m}^2 = 0.6\,\text{m}^2$

Average air velocity $= 0.45\,\text{m s}^{-1} = 1620\,\text{m h}^{-1}$

Hence volume of air entering room $= 1620\,\text{m h}^{-1} \times 0.6\,\text{m}^2 = 972\,\text{m}^3\,\text{h}^{-1}$

Therefore air change rate $= 972/39.2 = 24.80$ air changes $\text{h}^{-1}$

In terms of acceptability, this value is above the minimum requirement for 20 changes per hour.

## 14.3 Calculation of airborne contamination levels

### ROSAMUND M. BAIRD

*School of Pharmacy and Pharmacology, University of Bath, Claverton Down,*
*Bath BA2 7AY, United Kingdom*

Airborne contamination rates were monitored using a RCS Biotest (sampling rate $40\,\text{l min}^{-1}$) in a clean room over a 5-week period of time. A worked example of the mean airborne contamination level for each sampling point, expressed as the number of c.f.u. $\text{l}^{-1}$ or number of c.f.u. $\text{ft}^{-3}$, is calculated from the table below. Comments are given on the variability of the data and any follow-up action required.

Colony counts per 4-min sampling time

| Sample point | Week no. | | | | |
|---|---|---|---|---|---|
| | 1 | 2 | 3 | 4 | 5 |
| A | 39 | 36 | 33 | 21 | 98 |
| B | 17 | 56 | 25 | 14 | 72 |
| C | 18 | 34 | 20 | 8 | 57 |
| D | 26 | 15 | 48 | 12 | 63 |
| E | 7 | 23 | 31 | 25 | 50 |
| F | 19 | 44 | 29 | 44 | 67 |

From the above data, the following can be calculated using formulae in the Biotest product literature:

$$\text{Mean c.f.u.}\,l^{-1} = \frac{\text{No. of colonies}}{\text{Sampling rate} \times \text{Sampling time}} = \frac{\text{No. of colonies}}{40 \times 4}$$

$$\text{Mean c.f.u. ft}^{-3} = \frac{\text{No. of colonies} \times 0.708}{\text{Sampling time}}$$

Calculated contamination levels per unit volume

| Sampling point | Mean (c.f.u. $l^{-1}$) | Mean (c.f.u. $l^{-1}$) | Mean (c.f.u. ft$^{-3}$) |
| --- | --- | --- | --- |
| A | 45.4 | 0.284 | 8.036 |
| B | 36.8 | 0.23 | 6.514 |
| C | 27.4 | 0.171 | 4.85 |
| D | 32.8 | 0.205 | 5.806 |
| E | 27.2 | 0.17 | 4.814 |
| F | 40.6 | 0.254 | 7.186 |

The results show an overall increase in airborne counts during week 5 and should be reviewed against the set warning and action levels for this area. Other microbiological data (settle plates, contact plates and swabs) should also be reviewed for deviations in this area; the cleaning log should additionally be checked for any recorded problems.

## 14.4 Environmental monitoring in non-sterile manufacturing areas: a case study

MARTYN BECKER

*Medicines Control Agency, 3 East Grinstead House, London Road, East Grinstead, West Sussex, RH19 1RR, United Kingdom*

### 14.4.1 Introduction

This case study is developed from a number of actual cases, and it is intended to demonstrate the necessity for microbiological monitoring of the environment used to manufacture non-sterile dosage forms. It is not intended that the text should be used for the prescriptive setting of standards or microbiological targets; these will vary from environment to environment. The relevance of this example is that a manufacturing organization should be able to demonstrate control of its manufacturing environment.

### 14.4.2 *Background*

Two parties are involved: a company that submitted an application for a licence variation to allow contract manufacture of a preserved non-sterile cream, and a contract manufacturer; the two parties are identified as 'the company' and 'the contractor'. The company stated in the application that the contractor's manufacturing environment was grade D equivalent (Annex 1, EC GMP Guide) and that the manufacturing area was routinely monitored microbiologically.

### 14.4.3 *Inspection findings at the contractor's site*

Medicinal products were being manufactured alongside cosmetics in a large open processing area that was itself stated to be nominally graded D. Some of the vessels were uncovered. There is no current requirement that this type of manufacturing area is formally graded. Examination of environmental monitoring records indicated that levels of organisms in the background varied, up to $500 \, \mathrm{c.f.u. \, m^{-3}}$. It had also been noted that the preservative system in the cream was ineffective against two pigmented (and thus easily recognizable) bacterial species of the same genus that were normally present in the contractor's site environment. The company had addressed the problem by reformulating the cream, adding another preservative. Although the manufacturing background on the contractor's site was subject to routine microbiological monitoring, there were no set limits or action plan.

The contractor's site had begun working three shifts around the clock to fulfil orders. Trend analysis of monitoring data showed that levels of bacteria in the environment were increasing. Moreover, the two pigmented bacterial species noted during the original development work were *still routinely being detected*. The company representative stated that this was not a problem, as the amended preservative system would take care of any bacteria that happened to enter the product during manufacture.

The cream was indicated for use on open wounds.

Was this situation acceptable?

### 14.4.4 *Assessment*

1  The manufacturing area, although nominally grade D, was not segregated, such that pharmaceutical and non-pharmaceutical products could be manufactured side-by-side, thus presenting a cross-contamination risk.

2  Environmental monitoring was being undertaken at the contractor's site. There was, however, no limit applied because no work had been undertaken to characterize the normal background flora of the manufacturing environment. There was no system to assess the results and their implications, and feed-back any concerns to the manufacturing personnel.

3  Two organisms of the 'normal' flora had been demonstrated to withstand the original preservative system. There had been no programme to investigate the source of these organisms, whether they were acceptable in the manufacturing environment for this product, or how to remove them if they were not. These organisms were observed still to be routinely present in the manufacturing area.

4   The environment had not been demonstrated to be in control. The acceptability of the currently detected bacterial levels had not been justified. These levels were increasing, with no investigation planned to determine why, or what impact the increase to three manufacturing shifts was having. This was also important in that the product was indicated for use on open wounds and that some of the manufacturing vessels were uncovered.

5   The company's statement that the preservative system would remove any bacteria that entered the product during manufacture was inappropriate. The preservative helps to maintain the shelf-life of the product once opened in the market-place; in any case, two of the organisms detected had survived the original preservative formulation. The presence of preservative cannot be used to justify bad manufacturing practice.

### 14.4.5   *Summary*

Microbiological monitoring of non-sterile manufacturing areas is important from the perspective of controlling the manufacturing environment and thus the condition of the manufactured medicinal products. This can only be achieved if a system is in place to ensure that:

1   Good Manufacturing Practice is followed within an appropriate manufacturing environment and there is a set policy for manufacture regardless of the state of product preservation.

2   There has been an assessment of contamination risk with regard to the dosage form in question.

3   The position taken on microbiological monitoring is formally justified.

4   There has been a documented study carried out to characterize and quantify the 'normal' flora of the manufacturing environment, and this is acceptable for the product type being manufactured.

5   Limits are set against this normal background and unusual results or unexpected organisms investigated.

6   There has been a formal demonstration of the appropriateness of growth media used, sampling frequency and methods, locations, exposure times and incubation conditions.

7   The data derived from monitoring are analysed for trends and action taken in the event of an out-of-specification situation. This action should be seen to be effective.

## 14.5   Questions relevant to an audit of a culture medium manufacturer

ANDREW BILL

*Medicines Control Agency, 2nd Floor, Prudential House, 28-40 Blossom Street, York, YO24 1GJ, United Kingdom*

1   What assurance does the manufacturer have that the broth ingredients are of consistent quality, contain the required chemicals and do not contain impurities?

2  Are stocks of ingredients stored at the correct temperature, do they degrade with time, and how is expiry controlled?

3  Does the process use shared equipment? If so, is it cleaned adequately to remove traces of inhibitory agents from selective media, and how is this proven?

4  What quality of water is used. Is it suitable, and is it controlled?

5  Are the bottles clean before use, do closures stop micro-organisms gaining access during and after autoclaving, and can closures leach out inhibitory substances?

6  What in-process controls are used?

7  Is the autoclave cycle validated and how is that carried out?

8  Is proper consideration given to the risk that media may be subjected to excessive heat history with the risk of degradation of ingredients? Can media be re-autoclaved?

9  Is the autoclave properly maintained?

10  What system failures have occurred and what corrective actions have been put into place?

11  Is the process properly specified in written procedures?

12  What training and qualifications do the staff have?

13  How is the finished product released? What quality control checks are carried out on media?

14  What customer complaints are received?

15  Is mis-labelling possible?

16  Is the factory clean and suitable?

## 14.6  Calculation of linear regression equation

**ALAN BAINES**

*Managing Director, BioWhittaker UK Ltd, Wokingham, Berkshire,*
*RG41 2PL, United Kingdom*

This example illustrates the method of calculating the equation which represents a linear calibration plot; the endotoxin assay data in Table 9.4, Chapter 9 are used and summarized below. It is assumed that a 'scientific' calculator is available.

| Endotoxin source | Mean $\Delta$Absorbance (sample Absorbance – blank Absorbance) |
|---|---|
| 0.10 EU ml$^{-1}$ standard | 0.088 |
| 0.25 EU ml$^{-1}$ standard | 0.235 |
| 0.50 EU ml$^{-1}$ standard | 0.482 |
| 1.00 EU ml$^{-1}$ standard | 0.950 |
| Product #1 | 0.300 |
| Product #2 | 0.832 |

The following equations are used.

$$\text{Slope} = (Sy/Sx) \cdot r$$

$$y\text{-intercept} = \sum y/N - \left[\sum x/N \cdot \text{slope}\right]$$

$$r = \frac{N\sum xy - \sum x \cdot \sum y}{N(N-1)Sx \cdot Sy}$$

$$\text{Endotoxin concentration} = \frac{\text{O.D.} - y\text{-intercept}}{\text{Slope}}$$

where:

$x$ $\quad$ = Endotoxin concentration in EU ml$^{-1}$

$y$ $\quad$ = $\Delta$Absorbance value

$N$ $\quad$ = Number of standards used

$\sum x$ = Summation of concentration of standards used in EU ml$^{-1}$

$\sum y$ = Summation of $\Delta$Absorbance values

$\sum xy$ = Summation of the standard concentrations multiplied by the respective $\Delta$Absorbance values

$Sx$ $\quad$ = Standard deviation of $x$, given by the square root of:

$$\frac{N \cdot \sum x^2 - \left(\sum x\right)^2}{N(N-1)}$$

$Sy$ $\quad$ = Standard deviation of $y$ given by the square root of:

$$\frac{N \cdot \sum y^2 - \left(\sum y\right)^2}{N(N-1)}$$

$N$ $\quad$ = 4

$\sum x$ $\quad$ = 1.85 = $(0.1 + 0.25 + 0.5 + 1.0)$

$\sum y$ $\quad$ = 1.76 = $(0.088 + 0.235 + 0.482 + 0.95)$

$\sum xy$ = 1.26 = $(0.1 \times 0.088) + (0.25 \times 0.235) + (0.5 \times 0.482) + (1.0 \times 0.95)$

$Sx$ $\quad$ = 0.394

$Sy$ $\quad$ = 0.378

$$r = \frac{4(1.26) - (1.85)(1.76)}{4(4-1)(0.394)(0.378)} = 1.00$$

$$\text{Slope} = \frac{0.378}{0.394} \times 1.00 = 0.959$$

$$y\text{-intercept} = \frac{1.76}{4} - \left(\frac{1.85}{4} \times 0.959\right) = 0.439 - (0.462 \times 0.959) = -0.004$$

Product #1

$$\text{Endotoxin concentration (EU ml}^{-1}) = \frac{0.300 - (-0.004)}{0.959} = \frac{0.304}{0.959}$$

$$= 0.317 \, \text{EU ml}^{-1}$$

Product #2

$$\text{Endotoxin concentration (EU ml}^{-1}) = \frac{0.832 - (-0.004)}{0.959} = \frac{0.836}{0.959}$$

$$= 0.872 \, \text{EU ml}^{-1}$$

## 14.7 Relationship between inactivation factor, sterility assurance level and $F_0$ value

### NORMAN HODGES

*School of Pharmacy and Biomolecular Sciences, Brighton University, BN2 4GJ, United Kingdom*

For a terminally sterilized product pharmacopoeias require a minimum sterility assurance level (SAL) of $10^{-6}$. This means that the probability of an item in the sterilized batch remaining contaminated should be no greater than $10^{-6}$, or expressed another way, no more than one item in a million should contain a surviving organism.

The probability of survivors is influenced by the bioburden (the concentration of micro-organisms immediately prior to sterilization) and the inactivation factor (the extent to which a given sterilization cycle reduces the count). The inactivation factor (IF) is given by:

$$\text{IF} = 10^{t/D}$$

where $t$ = exposure time of the sterilization process and $D$ = the D value (decimal reduction time) of the bioburden organisms at the exposure temperature.

Assuming a 'worst-case scenario' in which the bioburden is $10^2$ c.f.u. ml$^{-1}$ and these are exclusively thermophilic bacterial spores possessing a heat resistance similar to that of *B. stearothermophilus* ($D_{121} = 1.6$ min).

For a 15-min/121°C autoclave cycle the inactivation factor would be

$$\text{IF} = 10^{15/1.6} = 10^{9.375} = 2.37 \times 10^9$$

The probability of survivors is given by:

$$P = \frac{\text{Bioburden}}{\text{Inactivation factor}} = \frac{10^2}{2.37 \times 10^9} = 4.22 \times 10^{-8}$$

Since this value is substantially lower than $10^{-6}$, the pharmacopoeial SAL is achieved.

It is possible to calculate the minimum exposure time required to achieve this SAL from the equation:

$$F_0 = D_{121} \times (\log N_0 - \log N_t)$$

where $F_0$ = the number of minutes exposure at 121°C which would have the same lethal effect as an alternative designated autoclave cycle, $N_0$ = the initial contamination level (bioburden) and $N_t$ = the final or required contamination level (SAL) after $t$ minutes exposure.

$$F_0 = 1.6 \times (\log 100 - (-6)) = 1.6 \times 8 = 12.8 \text{ min.}$$

## 14.8   Biological indicator case study

**NORMAN HODGES**

*School of Pharmacy and Biomolecular Sciences, Brighton University, BN2 4GJ, United Kingdom*

### 14.8.1   *The problem*

A company was commissioning a new autoclave and wished to use biological indicators (BIs) for revalidation of a 115°C/30-min sterilization cycle for an established product. The BIs available were aqueous spore suspensions of *B. stearothermophilus* spores sealed in glass ampoules and designed for use in a 121°C/15-min cycle. Their specification was:

Concentration of spores in the ampoule: $6.9 \times 10^4$ spores in 2 ml

$D_{121} = 1 \text{ min } 40 \text{ s}$

$Z$ (the number of °C temperature change required to produce a ten-fold change in D value) = 10.7°C

When these BIs were used in the validation study a significant proportion showed survivors after exposure. The company was confident that this was not due to poor operator aseptic technique, but suspected that the heat resistance of the spores in the BI was higher than the label claim. This resulted in a proposal to commission an independent determination of the $D_{121}$ value to ascertain whether this was so.

### 14.8.2   *The solution*

A theoretical analysis of the BI specifications was performed and this revealed that the observed spore survival was a result that might have been expected anyway, irrespective of the precise conformation of the BI to the label claim. The theoretical treatment leading to this conclusion is summarized below.

The $Z$ value is determined from the following equation:

$$Z = \frac{T_2 - T_1}{\log D_1 - \log D_2}$$

where $D_2$ and $D_1$ are the decimal reduction times at temperatures $T_2$ and $T_1$, respectively.

Thus, it is possible to calculate the $D_{115}$ value for the spores since $Z = 10.7$; $T_1 = 115°C$; $T_2 = 121°C$; $D_1 = D_{115}$; and $D_2 = D_{121}$.

The $D_{121}$ value of 1 min 40 s (1.67 min) has a log of 0.222

$$10.7 = \frac{121 - 115}{\log D_1 - 0.222}$$

$$\log D_1 = \frac{(121 - 115)}{10.7} - 0.222$$

$$D_{115} = 6.06 \text{ min.}$$

The theoretical number of surviving spores in each ampoule after 30 min exposure at 115°C is calculated from:

$$D = \frac{t}{\log N_0 - \log N_t}$$

where $D = 6.06$; $t = 30$ and $\log N_0 = \log$ of initial number of spores per ampoule ($\log 6.9 \times 10^4 = 4.839$); and $\log N_t = \log$ number of surviving spores per ampoule after 30 min exposure.

$$6.06 = \frac{30}{4.839 - \log N_t}$$

Rearranging the equation and calculating for $N_t$, the mean number of surviving spores per ampoule after 30 min at 115°C is

$$\log N_t = \frac{30}{6.06} - 4.839 = -0.1118$$

$$N_t = 0.773$$

Given the mean number of spores per ampoule, it is possible to calculate, using the Poisson equation, the proportions of ampoules containing zero survivors.

$$P = e^{-m}$$

where $P$ is the probability or frequency of occurrence of ampoules containing zero surviving spores; and $m$ is the *mean* number of surviving spores per ampoule. Thus, the proportion of ampoules containing zero surviving spores is:

$$P = e^{-0.773} = 0.46$$

### 14.8.3  *Conclusion*

Theoretically, 46% of exposed BIs would contain no survivors, but the majority, 54%, would contain one or more surviving spores. Hence, there was no point in conducting any D-value determinations since it could be predicted that just over one-half the exposed ampoules would be positive. The fundamental cause of the problem in this case study is that the characteristics of the BI rendered them unsuitable for their intended use in the first place.

# Index